Praise for *The Road Ahead*

"Even David Letterman boasted to Mr. Gates of how little he knows about computers. If he and the millions of Americans like him were to actually read *The Road Ahead*, they would discover that digital ignorance offers no protection from a future that will arrive whether we want it to or not."
　　—Frank Rich, *The New York Times*

"*The Road Ahead* does a remarkably good job of explaining the fundamentals of the computer revolution to people who remain hazy about what has really happened."
　　—James Fallows, *The New York Review of Books*

"An excellent review of the technology landscape."
　　—*Los Angeles Times*

"Bill Gates paints a dazzling portrait of the future. . . . People are bound to be entranced by this Asimovian vision of the information age."
　　—*Entertainment Weekly*

"This book gives us the more interesting—and more—Bill Gates, the man who always stood apart from his fellow technophiles because of his relentless focus on the bottom line. . . . It is a hardheaded look at what exactly it will take to wire the 21st-century world. . . . The technically challenged will find this . . . the best and most accessible writing on these subjects anywhere."
　　—Fred Moody, author of *I Sing the Body Electronic,* in the *Seattle Weekly*

"Fascinating . . . Gates's boundless optimism resounds between the covers of this book with an evangelical fury."
　　—*New York Daily News*

"Will undoubtedly advance the discussion about the digital age."
　　—*San Franscisco Chronicle*

"Since the mid-1970s, no one has had a clearer picture of how computing would evolve than Mr. Gates. Luckily, he sees a wondrous trip for us on the information highway."
　　—*The Wall Street Journal*

D0111361

PENGUIN BOOKS

THE ROAD AHEAD

BILL GATES is Chairman and Chief Executive Officer of Microsoft Corporation. Led by the belief that personal computers would be on every desktop and in every home, he cofounded Microsoft in 1975. His enduring vision and his goal of making software easier and more enjoyable for people to use are central to Microsoft's success and continue to shape the entire software industry. Born and raised in Seattle, Bill still resides in the Seattle area, with his wife, Melinda.

NATHAN MYHRVOLD, Ph.D., is Group Vice President, Applications & Content Group, at Microsoft Corporation. He joined Microsoft in 1986, when it acquired Dynamical Systems, a software company he had founded. Nathan has also held a position at Cambridge University working with Stephen Hawking. He earned a doctorate in theoretical/mathematical physics from Princeton University. Nathan recently served on the U.S. Advisory Council on the National Information Infrastructure.

PETER RINEARSON is a Pulitzer Prize–winning journalist who in 1982 wrote the first major profile of Bill Gates and Microsoft. He later wrote leading books on how to use Microsoft Word. In 1988 Peter founded Alki Software Corporation (alki.com), which publishes software utilities, Word add-ons, and Web content. He is also president of Raster Ranch, Ltd. (rasterranch.com), a digital production studio specializing in 3D design for the broadcast, game, and Internet/multimedia industries.

BILL GATES

WITH NATHAN MYHRVOLD
AND PETER RINEARSON

THE
ROAD
AHEAD

COMPLETELY REVISED
AND UP–TO–DATE

PENGUIN BOOKS

PENGUIN BOOKS

Published by the Penguin Group
Penguin Books Ltd, 27 Wrights Lane, London W8 5TZ, England
Penguin Books USA Inc., 375 Hudson Street, New York, New York 10014, USA
Penguin Books Australia Ltd, Ringwood, Victoria, Australia
Penguin Books Canada Ltd, 10 Alcorn Avenue, Toronto, Ontario, Canada M4V 3B2
Penguin Books (NZ) Ltd, 182–190 Wairau Road, Auckland 10, New Zealand

Penguin Books Ltd, Registered Offices: Harmondsworth, Middlesex, England

First published in Great Britain by Viking 1995
Published simultaneously in the USA by Viking Penguin
This revised edition published in Penguin Books 1996
1 3 5 7 9 10 8 6 4 2

Illustration credits:
Page 3: Courtesy of Lakeside School; Page 13: Reproduced by permission of Intel
Corporation, copyright 1972 Intel Corporation/Illustration courtesy of Microsoft Press;
Page 16: Reproduced from *Popular Electronics*, January 1975, copyright © 1975 Ziff-Davis
Publishing Company; Pages 27, 28, 78, 82, and 253: Illustrations courtesy of Microsoft
Press; Page 30: UPI/Bettmann; Page 35: Reproduced by permission of Intel Corporation,
copyright 1994 Intel Corporation; Page 53: Courtesy of International Business Machines
Corporation; Pages 57, 58, 135, 159, 164, 166, 221, and 275: Microsoft Corporation; Page
60: Courtesy of Apple Computer, Inc.; Page 73: Copyright 1995 Davis Freeman, Seattle,
WA; Page 80: Courtesy of Digital Equipment Corporation/Illustration courtesy of
Microsoft Press; Pages 249 and 250: Courtesy of Intergraph Corporation

Printed in England by Clays Ltd, St Ives plc

To my parents

PREFACE TO THE SECOND EDITION

I work in the software industry, where change is the norm. A popular software title, whether it's an electronic encyclopedia, a word processor, or an on-line banking system, gets upgraded every year or two with major new features and countless refinements. We listen to customer feedback and study new technology opportunities to determine the improvements to make.

For this new edition of *The Road Ahead*, I took the same approach. I made quite a number of changes, reviewing every paragraph and updating facts, but the biggest change is that the book, like Microsoft itself, now makes the Internet its central focus. The exploding popularity of the Internet is the starting point for most of the new material in this edition—including the total overhaul of chapters 5, 7, 9, and 11 and the afterword.

In the first edition, I said that the connection of personal computers and other information appliances would create a communications revolution. I'm surprised by how quickly it's happening and by the way it's happening. Although I used the early Internet as a student in the 1970s, I didn't expect then that the Internet's protocols would become the standard for a network everybody would be talking about twenty years later.

Even before the first PC was developed, I was enthusiastic about the potential for global networks that would connect the millions of personal computers I predicted would be "on every desk and in every home." I assumed that someday even home computers would be connected together, able to communicate with each other and draw on the world's pools of information. During the 1980s and early 1990s I was surprised that on-line services weren't very popular. I wondered what combination of network bandwidth and content it would take for electronic communication to become mainstream.

On a recruiting trip to Cornell University in late 1993, my technical assistant, Steven Sinofsky, was impressed by the way the academic community there was using the Internet to communicate. It wasn't just computer science students who were involved. Cornell and other universities were using the Internet to publish course schedules, student and faculty directories, class mailing lists and assignments, and news of lectures, exhibits and other events. High-speed networks linked personal computers that were available to every student. Clubs and organizations were conducting a lot of their business over the Internet. Students could even check their tuition bills over the net. Every student had an electronic mail (e-mail) address. Internet usage in the academic community had achieved critical mass. When I heard Steve talk about what was happening at Cornell, I began to take the Internet quite seriously.

By the spring of 1994 Microsoft was betting that the Internet would be important someday, and we were building support for it into our products. In addition we were spending more than $100 million annually for research and development on interactive networks of various kinds. But we didn't expect that within two years the Internet would captivate the whole industry and the public's imagination. We thought that relatively few people would be interested in pervasive interactivity until the technology supported videoconferencing and high-bandwidth applications such as video-on-demand—to say nothing of the needs for security, privacy, reliability, and convenience. We were great optimists in the long term, but the years of waiting for on-line services to catch on had made us conservative in our estimate of how soon significant numbers of people would be using interactive networks.

When the Internet really took off, we were surprised, fascinated, and pleased. Seemingly overnight people by the millions went onto the Internet, demonstrating that they would endure a lot more in the way of shortcomings than we had expected. People complained about the Internet's irritating deficiencies, but that didn't stop them from using this exciting new way to communicate. It was too much fun to ignore! All it had taken was for modems to get fast enough, communications switches to get cheap enough, PCs to get popular and powerful enough, and content on the Internet's World Wide Web to get rich enough, and there was no turning back. I can't tell you exactly when this point-of-no-return was reached, but by late 1995 we had crossed the threshold. More users meant more content and more content meant more users. The Internet had spiraled up in popularity, achieving critical mass.

Ironically, when a technology reaches critical mass its weaknesses and limitations almost become strengths as numerous companies, each trying to stake a claim in what quickly turns into a gold rush, step forward to fix the deficiencies. The original IBM PC is a good example of this phenomenon. The PC had arbitrary limitations that were easy to identify. Engineers at a lot of companies took one look at it and said, "Wow, that machine has problems!" When they rushed in to make money by remedying the PC's shortcomings, they set off the spiraling investment cycle that has driven the evolution of the PC's architecture for fifteen years. The Internet is in a similar position today. A lot of its pieces are missing or deficient, but because it is destined to evolve into the global information highway we've been talking about, the Internet offers a wonderful opportunity for companies who come up with ways to improve it.

The Internet is in an even stronger position than the PC was fifteen years ago. Far more people are contributing to its improvement than ever helped improve the PC. The pace of its evolution is so fast that the Internet is different every few months. Even those people most closely associated with it have to be impressed by how far the Internet has come, and how fast.

Today my company is among the thousands contributing to the Internet's evolution. It's no exaggeration to say that virtually everything Microsoft does these days is focused in one way or another on the Internet.

Signs of the Internet's success seem to be everywhere. TV ads run Web site addresses, and many people have at least one e-mail address on their business cards. Thousands of new pages go up on the Internet's World Wide Web every day. A panhandler in Seattle told me to check out his Web site. I thought, Boy, this is really getting popular! Maybe it was only a line, but I was impressed enough to give him what he wanted.

The level of investment in the Internet is amazing given that no one's making much profit yet. As some stock prices decline to more realistic levels, critics are sure to say that the Internet is nothing more than hype, or even dead.

Don't believe it. We're experiencing the early days of a revolution in communications that will be long-lived and widespread. There will be some surprises before we get to the ultimate realization of the information highway because much is still unclear. We don't understand consumer preferences yet. The role of government is a troubling open question. We can't anticipate all of the technical breakthroughs that lie ahead. But inter-active networking is here to stay, and it's only just beginning.

My thanks to Peter Rinearson, my chief collaborator for this edition, and to Erin O'Connor, our editor. My colleague Nathan Myhrvold pro-vided invaluable insight and guidance, as always. Craig Mundie consulted with me on hardware and infrastructure issues, and Tom Corddry on education issues. Jonathan Lazarus and Kelli Jerome coordinated the review of the revised manuscript and supervised revision of the CD-ROM. My thanks to all of them for their dedication.

For reviewing the revised chapters under a tight deadline, my thanks to Kimberly Ellwanger, Bob Gomulkiewicz, Tren Griffen, Jonathan Lazarus, Roger McNamee, Rick Rashid, and Steven Sinofsky.

At Microsoft Press, Elton Welke helped manage the project, and Buck Guderian, Michael Victor, and Bill Teel prepared new graphics. Thanks also to Peter Mayer, Pam Dorman, Susan VanOmmeren, and the other talented people at my English-language publisher, Viking Pen-guin, for their professionalism and patience.

It took the combined efforts of all of these people, and many others, to bring you this new edition of *The Road Ahead*. I hope you enjoy it and find it useful as you choose your own road in the years ahead.

—Bill Gates
July 25, 1996

FOREWORD

The past twenty years have been an incredible adventure for me. It all started on a day when, as a college sophomore, I stood in Harvard Square with my friend Paul Allen and pored over the description of a kit computer in *Popular Electronics* magazine. As we read excitedly about the first truly personal computer, Paul and I didn't know exactly how it would be used, but we were sure it would change us and the world of computing. We were right. The personal computer revolution has affected millions of lives. It has led us to places we barely imagined then.

We are all beginning another great journey. We can't be sure exactly where this one will lead either, but I'm certain it will touch many lives and take us all even farther. The major changes will be in how people communicate with each other. The benefits and problems arising from this upcoming communications revolution—which in its early stages we might call the "Internet Revolution"—will be much greater than those brought about by the PC revolution.

There is never a reliable map for unexplored territory, but we can learn important lessons from the creation and evolution of the $120 billion personal computer industry. After all, the PC—its evolving hardware, business applications, on-line systems, Internet connections, electronic mail, multimedia titles, authoring tools, and games—is the foundation for the next revolution.

During the PC industry's infancy, the mass media paid little attention to what was going on in the brand-new business. Those of us who were enthralled by computers and their possibilities went unnoticed outside our own circles. We were hardly what you'd call trendy.

But this next journey, on the so-called information highway, is the topic of an endless stream of newspaper and magazine articles, television and radio broadcasts, conferences, and rampant speculation. There has been an unbelievable amount of interest in the communications revolution during the last few years, both inside and outside the computer industry. The interest isn't confined to developed countries, and it goes well beyond even the very large numbers of personal computer users.

Some people think that the information highway—also called the interactive network—is simply the current version of the Internet or the delivery of hundreds of simultaneous channels of television. But today's innovations are just the beginning. The revolution in communications will take place over several decades and will be driven by new "applications"—new tools often meeting needs we don't even foresee now. Today's Internet only hints at tomorrow's.

During the next few years, governments, companies, and individuals will make major decisions about the network. It's crucial that broad groups of people—not just technologists or people who happen to be in the computer industry—participate in the debate about how this technology should be shaped and how it will in turn shape society.

I'm writing this book now as part of my contribution to the debate, and I hope it will stimulate more debate and inform the decisions we make. I do this with some trepidation. We've all smiled at predictions from the past that look silly today. You can flip through old *Popular Science* magazines and read about conveniences to come like the family helicopter and nuclear power "too cheap to meter." Then there was the Oxford professor who in 1878 dismissed the electric light as a gimmick and the professor who speculated in 1942 that the only way an airplane could break the sound barrier were if its wings flapped after it had been catapulted into the air. This is meant to be a serious book, although ten years from now it may not appear that way. What I've said that turned out to be right will be considered obvious and what was wrong will be humorous.

I believe the communications revolution will mirror in many ways the personal computer revolution. That's why in this book I draw on the history of the computing industry and on my own history for what they can teach us. Yes, I even talk about my house. But anyone expecting an autobiography or an account of what it's like to have been as lucky as I have been will be disappointed. When I've retired, I might get around to writing that book. This one looks primarily to the future.

Anybody expecting a technological treatise will be disappointed too. Everyone will be touched by the communications revolution, and everyone deserves to have its broad implications discussed in an accessible way. My goal from the very beginning was to write a book that as many people as possible would want to read.

Thinking about and writing *The Road Ahead* took longer than I thought it would. Estimating the time it would take proved to be as difficult as projecting the development schedule of a major software project. Even with able help from Peter Rinearson and Nathan Myhrvold, this book was a major undertaking. The only part that was easy was the cover photo by Annie Leibovitz, which we finished well ahead of schedule. I enjoy writing speeches, and I imagined in all innocence that writing a chapter would be equivalent to writing a speech. The fallacy in my thinking was a lot like the one that catches up software developers. A program ten times as long is actually about one hundred times more complicated to write. I should have known better. To complete the book, I had to take time off and isolate myself in my summer cabin with my PC.

And here it is. I hope it stimulates understanding, debate, and creative ideas about how we can take advantage of all that's sure to be happening in the decade ahead.

- ->

ACKNOWLEDGMENTS

Bringing a major software project to market can require the combined talents of hundreds of people. Not quite that many helped me with this book, but I certainly couldn't have done it alone. If I've inadvertently left someone out below, I'm really sorry, and I thank you too.

For everything from conception to marketing, and lots of stops along the way, thanks to Jonathan Lazarus and his team: Mary Engstrom, Kelli Jerome, Wendy Langen, and Debbie Walker. Without Jonathan's guidance and persistence this book never would have happened.

For their helpful suggestions throughout the project, special thanks to Tren Griffin, Roger McNamee, Melissa Waggener, and Ann Winblad.

For their incisive review comments, thanks to Stephen Arnold, Steve Ballmer, Harvey Berger, Paul Carroll, Steve Davis, Mike Delman, Kimberly Ellwanger, Brian Fleming, Bill Gates, Sr., Melinda Gates, Bernie Gifford, Bob Gomulkiewicz, Meg Greenfield, Collins Hemingway, Jack Hitt, Rita Jacobs, Erik Lacitis, Mich Mathews, Scott Miller, Craig Mundie, Daniel Petre, Rick Rashid, Jon Shirley, Mike Timpane, Wendy Wolf, Min Yee, and Mark Zbikowski.

For help with research, transcription, resource material, and other forms of support, my gratitude to Dale Askew, Kerry Carnahan, Ina Chang, Carol Gassner, Peggy Gunnoe, Michael Lee, Connie Lemmerman, Terri Lynch, Diana Murray, Patricia Riehl, Christine Shannon, Sean Sheridan, Denise Smith, Amy Dunn Stephenson, Susan Walsh, Laura Wright, and Virginia Wun. I'm also grateful to Elton Welke and his able staff at Microsoft Press, including Chris Banks, Judith Bloch, Jim Brown, Sally Brunsman, Mary DeJong, Jim Fuchs, Dail Magee, Jr., Erin O'Connor, JoAnne Woodcock, and Marc Young.

I'm also grateful to those at my English-language publisher, Viking Penguin, for their help and patience. In particular, I'd like to thank Peter Mayer, Marvin Brown, Barbara Grossman, Pamela Dorman, Cindy Achar, Kate Griggs, Theodora Rosenbaum, Susan Hans O'Connor, and Michael Hardart.

Thanks too, for editorial help, go to Nancy Nicholas and Nan Graham.

My special gratitude to my collaborators, Peter Rinearson and Nathan Myhrvold.

CONTENTS

A REVOLUTION BEGINS

I wrote my first software program when I was thirteen years old. It was for playing tic-tac-toe. The computer I used was huge and cumbersome and slow and absolutely compelling.

Cutting a bunch of teenagers loose on a computer was the idea of the Mothers' Club at Lakeside School in Seattle. The mothers decided that the proceeds from a big rummage sale should go to installing a terminal and buying computer time for the students. Letting students at a computer was a pretty progressive idea in the late 1960s—and a decision I'll always be grateful for.

The computer terminal didn't have a screen. To play, we typed in our moves on a typewriter-style keyboard and then sat around until the results came chug-chugging out of a loud printing device. We'd rush over to take a look and see who'd won or decide on a next move. A game of tic-tac-toe that would take thirty seconds with a pencil and paper might eat up most of a lunch period. But who cared? There was just something neat about the machine.

I realized later that part of the appeal must have been that here was an enormous, expensive, grown-up machine and we, the kids, could control it. We were too young to drive or do any of the other things adults could have fun at, but we could give this big machine orders and it would always obey.

Computers are great because when you're working with them you get immediate results: You know right away whether your program works. It's feedback you don't get from many other kinds of activity. The feedback from simple programs is particularly unambiguous. To this day it thrills me to know that if I can get the program right it will always work perfectly, every time, just the way I told it to. Experiencing this thrill was the beginning of my fascination with software.

As my friends and I got more confident, we started to mess around with the computer, speeding things up when we could or making the games more difficult. One of my friends at Lakeside developed a program in BASIC that simulated Monopoly play. BASIC (Beginner's All-purpose Symbolic Instruction Code), as its name suggests, is a relatively easy-to-learn programming language we used to develop increasingly complex programs. My friend figured out how to make the computer play hundreds of games really fast. We fed it instructions to test out various methods of play. We wanted to discover what strategies won most. And—chug-a-chug, chug-a-chug—the computer told us.

Like all kids, we not only fooled around with our toys, we changed them. If you've ever watched kids with a cardboard carton and a box of crayons create a spaceship with cool control panels, or listened to their improvised rules—"Red cars can jump all others"—you know that this impulse to make a toy do more is at the heart of innovation. It's the essence of creativity.

Of course, in those days we were just goofing around, or so we thought. But the toy we were goofing around with—well, it turned out to be some toy. A few of us at Lakeside wouldn't stop playing with it. In the minds of a lot of people at school we became identified with the computer, and it with us. A teacher asked me to help teach computer programming, and that seemed to be OK with everybody. But when I got the lead in the school play, *Black Comedy*, some kids were heard muttering, "Why did they pick the computer guy?" That's the way I still sometimes get labeled.

*1968: Bill Gates (standing) and Paul Allen working at the computer
terminal at Lakeside School*

A whole generation of us computer guys, all over the world, dragged
that favorite toy with us into adulthood. We caused a kind of revolu-
tion--peaceful, mainly—and now the computer has taken up residence
in our offices and in our homes. Computers have shrunk in size and
grown in power as they've dropped dramatically in price. And it's all
happened fairly quickly. Not as quickly as I once thought it would, but
still pretty fast. Inexpensive computer chips now show up in engines,
watches, antilock brakes, fax machines, elevators, gas pumps, cameras,
thermostats, treadmills, vending machines, burglar alarms, and even
talking greeting cards. School kids today are doing sophisticated things
with personal computers that are no bigger than textbooks but that out-
perform the largest computers of a generation ago.

Now that computing is astoundingly inexpensive and computers
inhabit every part of our lives, we stand at the brink of another revolution. This one will involve unprecedentedly inexpensive communication. All the computers will join together to communicate with us and
for us. Interconnected globally, they'll form a large interactive network,

which is sometimes called the information superhighway. The direct precursor of this network is the present-day Internet, which is evolving rapidly in the right direction. The reach and use of the emerging interactive network, its promise and perils, is the subject of this book.

Every aspect of what's about to happen seems exciting to me. When I was nineteen, I caught sight of the future and based my career on what I saw. I turned out to have been right. But the Bill Gates of nineteen was in a very different position from the position I'm in now. In those days, not only did I have all the self-assurance of a smart teenager, but also nobody was watching me. If I failed—so what? Today my position is much more like the position of the computer giants in the seventies, and I hope I've learned some lessons from them.

At one time I thought I might want to major in economics in college. I eventually changed my mind, but in a way my whole experience with the computer industry has been a series of economics lessons. I've been impressed firsthand by the effects of positive spirals and inflexible business models. I've watched the way industry standards evolved. I've seen the importance in technology of compatibility, of feedback, and of constant innovation.

I'm not using those lessons just for theorizing about this new era. I'm betting on it. Back when I was a teenager, I envisioned the impact of low-cost computers. I thought we could have "a computer on every desk and in every home," and that became Microsoft's corporate mission. We've helped to make that possible. Now those computers are being connected to each other, and we're building the software—the instructions that tell the computer hardware what to do—that will help people everywhere get the communication power of this connected universe. It's impossible to predict exactly what it will be like to use the fully realized broadband interactive network. It does look as if we'll communicate with the network through a variety of devices, including some that look like television sets, some like today's PCs, some like telephones, and some the size and something like the shape of a wallet. And at the heart of many of these devices will be a powerful computer, invisibly connected to millions of others.

The day has almost arrived when you can easily conduct business, study, explore the world and its cultures, call up great entertainment, make friends, go to neighborly markets, and show pictures to your relatives, wherever they are—without leaving your desk or your armchair. Once this new era is in full swing, you won't leave your network connection behind at the office or in the classroom. Your connection will be more than an appliance you've bought or an object you carry. It will be your passport into a new, "mediated" way of life.

Firsthand experiences are unmediated. No one will take away from you in the name of progress the experience of lying on a beach, walking in the woods, shopping at a flea market, or breaking up at a comedy club. But firsthand experiences aren't always so rewarding. Waiting in line is a firsthand experience, but we have been trying to invent ways to avoid it ever since we first queued up.

Tools are mediators, and much of human progress has come about because someone invented a better and more powerful tool. Physical tools speed up work and rescue people from hard labor. The plow and the wheel, the crane and the bulldozer, amplify the physical abilities of the people who use them.

Information tools are symbolic mediators. They amplify intellect rather than muscle. You're having a mediated experience as you read this book: We're not actually in the same room, but you're still able to find out what's on my mind. A great deal of our work now involves knowledge and decision-making, so information tools have become and will continue increasingly to be the focus of inventors. Just as any word can be represented by an arrangement of letters, information of all types can be represented in digital form, in a pattern of electrical pulses that's easy for computers to deal with. The world today has more than 100 million computers whose purpose is to help us manipulate information. Right now they help us by making it much easier to store and transmit information that's already in digital form, but in the near future they'll give us access to almost any information in the world.

In the United States, the connecting of the world's computers into an interactive network has been compared to another massive project: the interstate highway system, which got under way during the Eisenhower era. In the early 1990s the "information superhighway" seemed like an obvious choice of metaphor, and then-Senator Al Gore, whose father sponsored the 1956 Federal Aid Highway Act, popularized the term.

-------------------------------→

The highway metaphor isn't quite right, though. It suggests landscape, geography, a distance between points, and it implies that you have to travel, to move from one place to another. But in fact, this new communications technology will eliminate distance. It won't matter whether someone you contact is in the next room or on another continent because this highly mediated network will be unconstrained by miles and kilometers.

"Highway" also suggests that everybody is taking the same route. This network is more like a system of country roads. Everybody can take his own route, at his own speed, in his own direction. Another implication of the highway metaphor, perhaps, is that it should be built by the government—probably a major mistake in most countries. But the real problem with the highway metaphor is that it emphasizes the infrastructure rather than its applications. At Microsoft we talk about "information at your fingertips," which spotlights a benefit rather than the means.

A metaphor I prefer is the market. It comes closer to describing a lot of the activities that will take place on the network. The interactive network will be the ultimate market. Markets from trading floors to malls are fundamental to human society, and I believe this new one will eventually be a central place where we social animals will buy, sell, trade, invest, haggle, pick stuff up, argue, meet new people, and hang out. When you think of the interactive network, think of the hustle and bustle of the New York Stock Exchange or a farmers' market, or think of a bookstore full of people hunting down fascinating stories and information. All manner of human activity will take place on the network, from billion-dollar deals to flirtations. Many transactions will involve money, tendered in digital form rather than currency. Digital information of all kinds, not just as money, will be the medium of exchange in this market.

The global information market will combine all the various ways human goods, services, and ideas are exchanged. On a practical level, this will give you broader choices about most things, including how you earn and invest, what you buy and how much you pay for it, who your friends are and how you spend your time with them, and where and how securely you and your family live. Your workplace and your idea of what it means to be "educated" will be transformed, perhaps almost beyond recognition. Your sense of identity, of who you are and where you belong, may open up considerably. In short, just about everything will be done differently.

You aren't sure you believe this? Or want to believe it? Perhaps you'll decline to participate. People commonly take this position when a new technology threatens to change what they're familiar and comfortable with. At first the bicycle seemed like a silly contraption; the automobile, like a noisy intruder; the pocket calculator, a threat to the study of mathematics; and the radio, the end of literacy.

But vows to resist an innovation don't usually get kept. Over time, the new machine finds a place in our everyday lives because it not only offers convenience and saves labor, but it can also inspire us to new creative heights. It assumes a trusted place beside our other tools. A new generation grows up with it, changing and humanizing it—playing with it.

"Little by little, the machine will become a part of humanity," the French aviator and author Antoine de Saint-Exupéry wrote in his 1939 memoir, *Wind, Sand, and Stars.* He was describing how people tend to react to new technology, and he used the initial resistance to the railroad in the nineteenth century as an example. Saint-Exupéry pointed out that the smoke-belching, demonically loud engines of the primitive locomotives were called "iron monsters" at first. But as more tracks were laid, towns built train stations. Goods and services flowed. Interesting new jobs became available. A culture grew up around this novel form of transportation, and disdain became acceptance, even approval. What had been known as the iron monster became "the iron horse." "What is it today for the villager except a humble friend who calls every evening at six?" Saint-Exupéry asked.

The telephone was a major advance in two-way communication. But at first even it was denounced as a nuisance. People felt uncomfortable and awkward with this mechanical invader in the house. Eventually, though, they realized that they were not just getting a new machine; they were learning a new kind of communication. A talk on the phone didn't last as long and wasn't as formal as a face-to-face conversation. There was an unfamiliar and for many an off-putting efficiency to it. Before the phone, any good talk had entailed a visit and probably a meal, and people could expect to spend a full afternoon or an evening. Once most businesses and households had telephones, though, people came up with ways to take advantage of the unique characteristics of this means of communicating. As the telephone flourished, its own special expressions, tricks, etiquette, and culture developed. I'm sure Alexander Graham Bell never anticipated the silly executive game Have My Secretary Get Him Onto the Line Before Me. Now a newer form of communication—electronic mail, or e-mail—is undergoing the same process: establishing its own culture and conventions.

The single shift that has had the greatest effect on the history of communication took place about 1450, when Johann Gutenberg, a goldsmith from Mainz, Germany, invented movable type and introduced the first printing press to Europe. (China and Korea already had presses.) That event changed Western culture forever. Before Gutenberg all books were copied by hand. Monks, who usually did the copying, seldom managed more than one text a year. It took Gutenberg two years to compose the type for his first Bible, but once that was done, he could print multiple copies.

The printing press did more than just give the West a faster way to reproduce a book. Until it came on the scene, life had been communal and nearly unchanging despite the passing generations. Most people knew only what they had seen themselves or been told. Few people strayed far from their villages, in part because without reliable maps it was often nearly impossible to find the way home. As James Burke, a favorite author of mine, put it: "In this world all experience was personal: horizons were small, the community was inward-looking. What existed in the outside world was a matter of hearsay."

The printed word changed all that. It was the first mass medium. For the first time knowledge, experiences, and opinions could be passed on in a portable, durable, easily available form. As the written word extended the population's reach far beyond the village, people began to care about what was happening in the wider world. Printing shops sprang up in commercial cities and became centers of intellectual exchange. Literacy was a significant skill that revolutionized education and changed social structures.

Before Gutenberg there were only about 30,000 books on the entire continent of Europe, nearly all of them Bibles or biblical commentary. By 1500 there were more than 9 million books, on all sorts of topics. Handbills and other printed matter affected politics, religion, science, and literature. For the first time, people outside the canonical elite had access to written information.

The global interactive network will transform our culture as dramatically as Gutenberg's press did the Middle Ages.

Personal computers have already changed our work habits, but it is the evolving Internet that will really change our lives. As information machines are connected on the Internet, people, entertainment, and information services suddenly become accessible. As the Internet's popularity and capability increase, you'll be able to stay in touch with anyone, anywhere, who wants to stay in touch with you and to browse through any of thousands of sources of information, day or night. A little further along, you'll be able to answer your apartment intercom from your office or answer any mail from your home. Your misplaced or stolen camera will send you a message telling you exactly where it is, even if it's in a different city. Information that once was difficult to retrieve will be easier and easier to find:

Is my bus running on time?

Are there any accidents right now on the route I usually take to the office?

Does anyone want to trade his or her Thursday theater tickets for my Wednesday tickets?

What is my child's school-attendance record?

What's a good recipe for halibut?

Which store, anywhere, can deliver by tomorrow morning for the lowest price a wristwatch that can take my pulse?

What would somebody pay for my old Mustang convertible?

How is the hole in a needle manufactured?

Are my shirts at the laundry ready yet?

What's the cheapest way to subscribe to the *Wall Street Journal*?

What are the symptoms of a heart attack?

Was there any interesting testimony at the county courthouse today?

Do fish see in color?

What does the Champs-Élysées look like right now?

Where was I at 9:02 P.M. last Thursday?

Let's say you're thinking about trying a new restaurant and want to see its menu, the wine list, and the specials of the day. Maybe you're wondering what your favorite food reviewer said about it. You may also want to know what sanitation score the health department gave the place. If you're leery of the restaurant's neighborhood, you might want to see a safety rating based on police reports. Still interested in going? You'll want reservations, a map, and directions based on current traffic conditions. You'll take the directions in printed form or have them read to you—and updated—as you drive.

All of this information will be readily accessible and completely personalized for you. You'll be able to explore whatever parts of it interest you in whatever ways and for however long you want. You'll watch a program when it's convenient for you instead of when a broadcaster chooses to air it. You'll shop, order food, contact friends, or publish information for other people to use when and as you want to. Your nightly newscast will start at a time you determine and last exactly as long as you want it to, and it will cover subjects selected by you or by a service that knows your interests. You'll be able to ask for reports from Tokyo or Boston or Seattle, request more detail on a news item, or inquire whether your favorite columnist has commented on an event. If you prefer, your news will be delivered to you on paper.

Early forms of some of these services are showing up on the Internet already, but they only hint at what's to come. A massive shift in the way people communicate and relate to information is under way.

Change of this magnitude makes people nervous. Every day, all over the world, people are asking about the implications of information technology, often with apprehension. What will happen to our jobs? Will we withdraw from the physical world and live vicariously through our computers? Will the gulf between the haves and the have-nots widen irreparably? Will a computer be able to help the disenfranchised in East St. Louis or the starving in Ethiopia? Without a doubt, major challenges will accompany the network and the changes it will bring about. In chapter 12, I talk about many of the legitimate concerns I hear people express again and again.

Information technology is not a panacea. This disappoints people who demand to know how PCs and the Internet will solve all human problems. I wonder whether in Gutenberg's age people asked: "What good is this press? Is it going to feed people? Will it help overcome illness? Will it make the world more just?" Eventually it facilitated all of these things, of course, but in 1450 it was probably hard to tell that it would.

One thing is clear: We don't have the option of turning away from the future. No one gets to vote on whether technology is going to change our lives. No one can stop productive change in the long run because the marketplace inexorably embraces it. Governments can try to slow the rate of change within their own borders by restricting the use of certain technologies, but these policies risk leaving a country isolated from the world economy, preventing its companies from being competitive and its consumers from getting the latest products and the best prices.

I believe that because progress will come no matter what, we need to make the best of it—not try to forestall it.

I'm still thrilled by the feeling that I'm squinting into the future and catching that first revealing hint of revolutionary possibilities. I first experienced this particular kind of euphoria as a teenager, when I began to understand how inexpensive and powerful computers would become. The mainframe we played tic-tac-toe on in 1968, like most computers at that time, was a temperamental monster that lived in a climate-controlled cocoon. After we had used up the Mothers' Club grant, Paul Allen and I spent a lot of time trying to get access to computers. The computers performed modestly by today's standards, but they seemed awesome to us because they were big and complicated and cost as much as millions of dollars each. They were connected by phone lines to clackety Teletype terminals so that they could be shared by people at different locations. We rarely got close to the mainframes themselves.

Computer time was very expensive. When I was in high school, it cost about $40 an hour to access a time-shared computer using a Teletype—for that $40 an hour you got a slice of the computer's precious attention. This seems odd today, when some people own more than one PC and think nothing of leaving their computers idle for most of the day. Actually, it was possible even then to own your own computer. If you could afford $18,000, Digital Equipment Corporation (DEC) would sell you a PDP-8. Although it was called a "mini-computer," the PDP-8 was physically big by today's standards. It occupied a rack about two feet square and six feet high, and it weighed 250 pounds. We had one at our high school for a while, and I fooled around with it a lot. The PDP-8 was very limited compared to the mainframes we could reach by phone; it had less raw computing power than some wristwatches do today. Nevertheless, it was programmable the same way the big, expensive computers were: You gave it software instructions. Despite its limitations, the relatively inexpensive PDP-8 inspired us to indulge in the dream that one day millions of people would have their own computers. With each passing year, we became more certain that computers and computing would be cheap and widespread. I'm sure that one reason I was so determined to help develop the personal computer is that I wanted one for myself.

Back then software, like computer hardware, was expensive. It had to be written specifically for each kind of computer. And each time the computer hardware changed, which it did regularly, the software for it had to be pretty much rewritten. Computer manufacturers provided some standard software program building blocks with their machines (libraries of mathematical functions, for example), but most software had to be custom-written to solve some business's specific, unique problems. Some software could be shared, and a few companies sold general-purpose software, but there was very little packaged software you could buy off the shelf.

My parents paid my tuition at Lakeside and gave me money for books, but I had to pay my own computer-time bills. This is what drove me to the commercial side of the software business. I needed money to buy access. A bunch of us, including Paul Allen, got entry-level software programming jobs in the summers. For high school students, the pay was extraordinary—about $5,000 each summer, part in cash and the rest in computer time. We also worked out deals with a few companies whereby we could use computers for free if we'd identify problems in their software.

One of the early programs I wrote, not for money, was for Lakeside. It scheduled students in classes. I surreptitiously added a few instructions and found myself nearly the only guy in a class full of girls. It was hard to tear myself away from a machine at which I could so unambiguously demonstrate success. I was hooked.

Paul knew a lot more than I did about computer hardware, the machines themselves. One summer day in 1972, when I was sixteen and Paul was nineteen, he showed me a ten-paragraph article buried on page 143 of *Electronics* magazine. It announced that a young company called Intel had released a microprocessor chip they'd named the 8008.

A microprocessor is a simple chip that contains the entire brain of a computer. Paul and I realized that this first microprocessor was very limited, but he was sure that the chips would get more powerful, that computers on a chip would improve very rapidly. This insight of Paul's was the cornerstone of all that we did together later, including the founding of Microsoft.

At the time, the computer industry had no idea of building a real computer around a microprocessor. The *Electronics* article, for example, described the 8008 as suitable for "any arithmetic, control, or decision-making system, such as a smart terminal." The writers didn't see that a microprocessor could grow up to be a general-purpose computer. Microprocessors then were slow and limited in the amount of informa-

1972: Intel's 8008 microprocessor, about the size of the tip of a ballpoint pen

tion they could handle. And none of the languages that programmers were familiar with was available for programming instructions for the 8008. That made it nearly impossible to write a program of any complexity for it. Every application for the 8008 had to be programmed with the few dozen simple instructions the chip could understand. The 8008 was condemned to life as a beast of burden, carrying out uncomplicated and unchanging tasks over and over. It was quite popular in elevators and calculators.

To put it another way, a simple microprocessor in an embedded application such as an elevator's controls is a single instrument, a drum or a horn, in the hands of an amateur: good for basic rhythm or uncomplicated tunes. A more complex microprocessor for which there are programming languages, however, is like an accomplished orchestra. With the right score, or software, it can play anything.

Paul and I wondered what we could program the 8008 to do. Could we write a version of BASIC that would run on it? Paul asked. He called up Intel to see if he could get a manual. We were a little surprised when they actually sent him one. We both dug into it. I had worked out a version of BASIC that ran on the limited DEC PDP-8, and I was excited at the thought of doing the same for the little Intel chip. But as I studied the 8008's manual, I realized it was futile to try. The 8008 just wasn't sophisticated enough, didn't have enough transistors.

We did figure out a way to use the little chip to power a machine that could analyze information counted by traffic monitors on city streets. Many municipalities measured traffic flow by stringing a rubber hose over a selected street. When a car crossed the hose, it punched a paper tape inside a metal box at the end of the hose. We saw that we could use the 8008 to process these tapes, to print out statistics and graphs. We called our first company "Traf-O-Data." We thought that name was sheer poetry.

I wrote much of the software for the Traf-O-Data machine on cross-state bus trips from Seattle to Pullman, Washington, where Paul was going to college. (I was still at Lakeside.) Our prototype worked well, and we envisioned selling lots of our new machines all over the country. We used the prototype to process traffic-volume tapes for a few municipal customers, but no one actually wanted to buy the machine, at least not from a couple of teenagers.

We were disappointed, but we still believed that our future, even if it was not to be in hardware, might have something to do with microprocessors. After I started at Harvard University in 1973, Paul coaxed his clunky old Chrysler New Yorker across the country from Pullman and took a job in Boston, programming minicomputers at Honeywell. He drove over to Cambridge a lot so that we could keep talking about our schemes for the future.

In the spring of 1974, *Electronics* magazine announced Intel's new 8080 chip—with ten times the power of the 8008 inside our Traf-O-Data machine. The 8080 wasn't much bigger than the 8008, but it contained 2,700 more transistors. All of a sudden we were looking at the heart of a real computer, and the price was under $200. Paul wondered if this chip was powerful enough to support a version of BASIC. We attacked the manual and concluded that BASIC for the 8080 was feasible. "DEC won't be able to sell any more PDP-8s now," I told Paul. It seemed obvious to us that if a tiny chip could get so much more powerful inside of two years, the end of the big unwieldy machines was at hand.

Computer manufacturers didn't see the microprocessor as a threat, though. They just couldn't imagine a puny chip taking on a "real" computer. Not even the scientists at Intel saw their chip's full potential. To them, the 8080 represented nothing more than an improvement in chip technology. And in the short term the computer establishment was right. The 8080 was just another incremental advance. But Paul and I looked past the limits of that new chip and saw a different kind of computer that would be perfect for us, and for everybody—personal, affordable, and adaptable. It was absolutely clear to us that because the new chips were so cheap they would soon be everywhere.

We saw that computer hardware, which had once been scarce, would soon be readily available and that access to computers would no longer be subject to a high hourly charge. It seemed to us that people would find all kinds of new applications for computing if it was cheap. Then software would be the key to delivering on the full potential of these machines. Paul and I thought that Japanese companies and IBM would likely produce most of the hardware. We believed we could come up with new and innovative software. And why not? The microprocessor

------------------------------➤

*January 1975 issue of
Popular Electronics*

would change the structure of the industry. Maybe there was a place in the new scheme of things for the two of us.

This kind of talk is what the college years are all about. You have all kinds of new experiences, and you dream crazy dreams. We were young and assumed that we had all the time in the world. I enrolled for another year at Harvard and kept thinking about how we could get a software company going. We sent letters from my dorm room to all the big computer companies, offering to write them a version of BASIC for the new Intel chip. We got no takers. By December of 1974 we were pretty discouraged. I was planning to fly home to Seattle for the holidays, and Paul was going to stay on in Boston. On an achingly cold Massachusetts morning a few days before I left, Paul rushed me to the Harvard Square newsstand to show me the January issue of *Popular Electronics*.

On the magazine's cover was a photo of a very small computer, not much larger than a toaster oven. Its name was a little more resonant than Traf-O-Data: the Altair 8800. ("Altair" was the star orbited by the For-bidden Planet, and several planets that supposedly orbit Altair figure in various *Star Trek* episodes.) It was being sold as a kit for $397. When it was assembled, it had no keyboard or display—just sixteen address switches to direct commands, and sixteen lights. You could get the little lights on the front panel to blink, but that was about all. Part of the Altair 8800's problem was that it didn't have any software, which meant that it couldn't be programmed, and that made it more a novelty than a practical tool.

What the Altair did have was that Intel 8080 microprocessor chip as its brain. When we saw that, panic set in. "Oh no! It's happening without us! People are going to go write real software for this chip." The future was staring us in the face from the cover of a magazine. It wasn't going to wait for us. Getting in on the first stages of the PC revolution looked like the opportunity of a lifetime, and we seized it.

Twenty years later there are many parallels to our situation then but major differences too. An obvious difference is that while back then I was afraid other companies would scoop us, today I know that thousands of companies already share our vision of interactive networks. Literally millions of people are trying to take advantage of the opportunities afforded by the Internet. They know that the legacy of the earlier revolution is that 50 million PCs are sold each year worldwide and that the fortunes of companies in the computer industry were completely reordered in that revolution. Many people sense that that kind of change could happen again. So the rush is on to get in early while the opportunities seem infinite—and before competitors can get an upper hand.

When we look back at the last twenty years, it's apparent that some large computer companies were so set in their ways that they didn't adapt and lost out. The mistakes large companies made stemmed primarily from an inability to move quickly to seize new opportunity. When IBM failed to emphasize software and DEC failed to embrace the personal computer, they made room for the likes of Microsoft.

Twenty years from now we'll look back and see the same pattern. I know that as I write this there's at least one young person out there who will create a major new company, convinced that his or her insight into the communications revolution is the right one. Thousands of innovative companies will be founded to exploit the coming changes.

An early pioneer with that kind of insight is Marc Andreessen, one of

------------------------------>

the founders of Netscape, a software company formed in 1994 to ride the early currents of the Internet wave, much as Microsoft had been founded almost two decades earlier to catch the PC wave. Andreessen wants his company to be the next Microsoft, but the competitive climate is dramatically different for Netscape than it was for the early Microsoft. Microsoft was quite fortunate in its early years in that the established players of the industry all but ignored personal computing. We didn't face the kind of focused competition from competitors that Netscape will as it jockeys for a profitable leadership position. On the other hand, Netscape started with the advantage of experienced management and a high stock price that let it buy other companies.

In 1975, when Paul and I naively decided to start a company, we were like the characters in those Judy Garland and Mickey Rooney movies: "Let's put on a show in the barn!" We thought there was no time to waste, and we set right to it. Our first project was to create a version of BASIC for the little Altair computer.

We had to squeeze a lot of capability into the computer's small memory. The typical Altair had about 4,000 characters of memory. Today most new personal computers have at least 8 million characters of memory and often much more. Our task was further complicated because we didn't actually own an Altair—and had never even seen one. That didn't really matter because what we were really interested in was the new Intel 8080 microprocessor chip—and we'd never seen one of those either. Undaunted, Paul studied a manual for the chip and then wrote a program that made a big computer at Harvard mimic the little Altair. That gave us our "machine" on which to test our software. This was like having a whole orchestra available and using it to play a simple duet, but it worked.

Writing good software requires a lot of concentration, and writing BASIC for the Altair was exhausting. Sometimes I rock back and forth or pace when I'm thinking because it helps me focus on a single idea and exclude distractions. I did a lot of rocking and pacing in my dorm room the winter of 1975. Paul and I didn't sleep much and lost track of night and day. When I did fall asleep, it was usually at my desk or on the floor. Some days I didn't eat or see anyone. But after five weeks our BASIC was written—and the world's first microcomputer software company was born. In time we'd name it "Microsoft."

We knew that getting a company started would mean sacrifice. But we also realized that we had to do it then or forever lose the opportunity to make it in microcomputer software. In the spring of 1975 Paul quit his programming job and I decided to go on leave from Harvard.

I talked it over with my parents, both of whom were pretty savvy about business. My father, watching from Seattle, feared for my future. He told an old law school friend that his only son had talent but might not amount to anything. My mother was even more concerned. She composed a rhyme for the family Christmas card, a playful update on what my two sisters and I were up to. It expressed the hope that my software business was not a "turkey"—because its profit picture looked "murky." Despite their anxieties, my parents saw how much I wanted to start a software company and they were supportive. My plan was to take some time off, start the company, and then go back later and finish college. I never made a conscious decision to forgo a degree. Technically, I'm just on a really long leave.

I had loved college, both for the classes and for the students. When Harvard had asked us to describe our ideal roommates, I'd said that I'd like to live with a foreign student and a minority student. That's probably how a kid from Seattle, a Canadian, and an African-American from Tennessee ended up sharing a dorm room. College was fun because I could sit around and talk with so many smart people my own age. But I wasn't sure the window of opportunity for starting up a software company would open again, so I left and dove into the world of business when I was nineteen years old.

From the start Paul and I funded everything ourselves. Each of us had saved some money. Paul had been paid well at Honeywell, and some of the money I invested in our startup came from late-night poker games in the dorm. Fortunately, our company didn't require much funding.

People often ask me to explain Microsoft's success. They want to know the secret of getting from a two-man, shoestring operation to a company with more than 21,000 employees and more than $8 billion a year in sales. Of course, there's no simple answer, and luck has played a role, but I think the most important element was our original vision.

We thought we saw what lay beyond that Intel 8080 chip and then acted on it. We asked, "What if computing were nearly free?" We

------------------------------>

believed that there would be computers everywhere because computing power would be cheap and great new software would take advantage of it. We set up shop betting on cheap computer power and producing software when nobody else was. Our initial insight made everything else easier. We were in the right place at the right time. We got there first, and our early success gave us the chance to hire more and more smart people. We built a worldwide sales force and used the revenue it generated to fund new products. But from the beginning, we set off down a road that was headed in the right direction.

Now there's a new set of circumstances, and the relevant question this time is, "What if communicating were almost free?" The idea of interconnecting all those homes and offices to a high-speed interactive network has ignited imaginations around the world. Thousands of companies are committed to the same vision, so it's individual focus, a superior understanding of intermediate steps, and execution that will determine success.

All sorts of individuals and companies are betting their futures on building components for the interactive network. I call it the Internet Gold Rush. At Microsoft we're working hard to figure out how to evolve from where we are today to the point at which we can realize the full potential of numerous new advances in technology. These are exciting times, not only for the companies involved, but for everyone who will enjoy the benefits of this revolution.

2

THE BEGINNING OF
THE INFORMATION AGE

The first time I heard the term "Information Age," I was interested. I knew about the Iron Age and the Bronze Age, the periods of human history named for the metals men had discovered for making their tools and weapons. And of course, I'd learned about the Industrial Age in school. But when I read academic predictions that countries would fight over the control of information, not natural resources, I wasn't sure what they meant by information.

The claim that information would define the future reminded me of the famous party scene in the 1967 movie *The Graduate*. A businessman buttonholes Benjamin, the college graduate played by Dustin Hoffman, and offers him a single word of unsolicited career advice: "Plastics." I wondered whether, if the scene had been written a few decades later, the businessman's advice would have been: "One word, Benjamin. 'Information.'"

Maybe I was intrigued because information stood at the intersection of several of my interests in college. The kind of math I'm good at is called "combinatorics," which has among its practical applications the making and breaking of enciphered messages—secret codes. Using math to lock and unlock information really fascinated me. I was interested in economic game theory too, which uses math and logic to devise optimal competitive strategies. This got me to thinking about how information is valued in a contest in which each side keeps vital secrets. How much value can be assigned to information when more than one person has it? The value may fall to zero, especially if a price war breaks out between two or more people wanting to sell the same information.

It seemed to me that too many people were accepting at face value, uncritically, the idea that information was becoming the most valuable commodity. Information was at the library. Anybody could check it out for nothing. Didn't that accessibility undermine its value? And information could be wrong, in which case it might have negative value—it might hurt instead of help. Even when the information that bombarded us every day proved to be correct, most of it was irrelevant anyway. And when information was relevant, its value was often ephemeral, decaying with the passage of time or if too many people had it. A once-promising investment tip might become useless as it aged or if it became widely known. Yet information was held up as the central ingredient of an emerging world economy.

I could imagine nonsensical conversations around a future office watercooler: "How much information do you have?" "Switzerland is a great country because of all the information they have there!" "I hear the Information Price Index is going up!"

Talk like that sounds nonsensical because information isn't as tangible or measurable as the metals that defined some previous ages, but the academics were on to something. Information has become increasingly important to us, and indeed we're at the beginning of an information revolution. The cost of communications is beginning to drop, although not as precipitously as the cost of computing did. When communication gets inexpensive enough and is combined with other advances in technology, the influence of interactive information will be as real and as far-reaching as the effects of electricity.

To understand why information is getting to be so central, it's important to know how technology is changing the ways we handle information. The major part of this chapter is devoted to giving readers who aren't familiar with the history of computing and the principles by which computers handle information enough to go on so that they can enjoy the rest of the book. If you understand how digital computers work, feel free to skip ahead to chapter 3.

The most fundamental difference we'll see between information as we've known it and information in the future is that almost all information will be digital. Whole print libraries are already being scanned and stored as electronic data on disks and CD-ROMs. Newspapers and magazines are often composed completely in electronic form and printed on paper only as a convenience for distribution. Their electronic information is stored permanently—or for as long as anybody wants it—in computer databases, giant banks of journalistic data accessible through on-line services. Photographs, films, and videos are all being converted into digital information. Every year we come up with better methods for quantifying information and distilling it into quadrillions of atomistic bits of data. Once digital information is stored, anybody with permission and access through a personal computer can instantaneously recall, compare, and refashion it. What characterizes this period in history, what sets it apart, is that ability to refashion information—the completely new ways in which information can be manipulated and changed—and the increasing speeds at which we can handle information. The computer's ability to provide low-cost, high-speed processing and transmission of digital data will transform the conventional communication devices in our homes and offices.

The idea of using an instrument to manipulate numbers isn't new. Asians had been using the abacus for nearly 5,000 years by 1642, when the nineteen-year-old French scientist Blaise Pascal invented a mechanical calculator. It was a counting device. Thirty years later the German mathematician Gottfried von Leibniz improved on Pascal's design. His "Stepped Reckoner" could multiply, divide, and calculate square roots. Reliable mechanical calculators powered by rotating dials and gears, descendants of the Stepped Reckoner, were the mainstay of business operations right up until their electronic counterparts replaced them. When I was growing up, a cash register was essentially a mechanical calculator linked to a cash drawer.

More than a century and a half ago a visionary British mathematician

------------------------------▶

glimpsed the possibility of the computer, and that insight made him famous even in his day. Charles Babbage was a professor of mathematics at Cambridge University who conceived the possibility of a mechanical device that would be able to perform a string of related calculations. As early as the 1830s Babbage was drawn to the idea that information could be manipulated by a machine if the information could be converted into numbers first. The steam-powered machine Babbage envisioned would use pegs, toothed wheels, cylinders, and other mechanical parts, the apparatus of the then-new Industrial Age. Babbage believed that his "Analytical Engine" would take the drudgery and inaccuracy out of calculating.

Babbage didn't use the terms we use now to refer to the parts of his machine. He called the central processor, or working guts, of his machine "the mill." He referred to his machine's memory as "the store." Babbage imagined that information would be transformed the way cotton was—drawn from a store (warehouse) and milled into something new.

His Analytical Engine would be mechanical, but Babbage saw how it would be able to follow changing sets of instructions and thus serve different functions. And that's the essence of software, a comprehensive set of rules that tell a machine what to do, that "instruct" it, step by step, how to perform particular tasks. Babbage realized that he would need an entirely new kind of language for giving the machine these instructions, and he devised one that used numbers, letters, arrows, and other symbols. Babbage designed the language to enable him to "program" the Analytical Engine with a long series of conditional instructions that would allow the machine to modify its actions in response to changing situations. He was the first to see that a single machine could serve a number of different purposes—unlike a cotton gin, which was designed to do only one task over and over again. Babbage saw that a general-purpose machine, running software, could replace countless special-purpose machines.

For the next one hundred years mathematicians worked with ideas derived from Babbage's and by the mid-1940s had finally built an electronic computer based on the principles of his Analytical Engine. It's hard to sort out the paternity of this modern computer because much of the thinking and the actual building was done in the United States and

Britain during World War II, under the cloak of wartime secrecy. We do know that Alan Turing, Claude Shannon, and John von Neumann were three major contributors.

In the mid-1930s Alan Turing, like Babbage a superlative Cambridge-trained British mathematician, proposed what is known today as a "Turing machine." It was his version of a general-purpose calculating machine that could be instructed to work with almost any kind of information.

In the late 1930s, when Claude Shannon was still a student, he demonstrated how a machine executing logical instructions could manipulate information. Shannon's insight, the subject of his master's thesis, was that computer circuits—closed for true and open for false—could perform logical operations: the number 1 could represent "true," and 0 could represent "false."

This is a binary system, a code. Binary expression is the alphabet of electronic computers, the basis of the translation, storage, and manipulation of all information within a computer. Each 1 or 0 is a "bit" of information.

We had some fun with binary numbers in the backyard of my house in the summer of 1991, when several groups of friends competed with each other in a smoke-signal competition. The object was to encode information for rapid transmission. Each team was assigned a smoke machine, which two or three members of the team were to operate while everybody else on the team stood back to receive the signals, as in charades. Each team had twenty minutes to devise a code that would let them use puffs of smoke to communicate a number. None of the contestants knew what the number would be, or even how many digits it would contain, so the code a team came up with had to be flexible.

With more than a few computer programmers on each of the teams, it was no surprise that the winners used a binary scheme, converting each digit of the secret number into a 4-bit binary number. This meant that 0 would be sent as 0000, 1 as 0001, 2 as 0010, 3 as 0011, and so on. Any number can be expressed in this scheme, which is called "binary coded decimal." The winning team decided that a short puff signified 0, and a long puff meant 1. As I recall, the secret number was 709, which the winners translated into 0111-0000-1001 and sent as short-long-long-

long-short-short-short-short-long-short-short-long. Actually, though, binary coded decimal wasn't the most efficient way of communicating the number. If 709 had been treated as a single binary number rather than as three separate digits, it would have been transmitted as 1011000101—ten puffs of smoke to the winning team's twelve. The winning team simply ran its smoke machine very well.

The binary system is simple, but it's so vital to an understanding of the way computers work that it's worth pausing here to explain it more fully.

Imagine that you want to illuminate a room with as much as 250 watts of electric light and you want the lighting to be adjustable, from 0 watt of illumination (total darkness) to the full wattage. One way to accomplish this is with a rotating dimmer switch hooked up to a 250-watt bulb. To achieve complete darkness, you'd turn the knob fully counterclockwise to Off for 0 watt of light. For maximum brightness, you'd turn the knob fully clockwise for the entire 250 watts. For some illumination level in between, you'd turn the knob to an intermediate position.

This system is easy to use, but it has its limitations. If the knob is at an intermediate setting—if you're lowering the wattage for an intimate dinner, say—you can only guess what the lighting level is. You don't really know how many watts you're using or how to describe the setting in any precise way. Your setting/information is only approximate, which makes it hard to store and hard to reproduce.

What if you wanted to reproduce exactly the same level of lighting the next week? You could make a mark on the switch plate so that you'd know how far to turn the knob, but that would hardly be exact, and what would happen when you wanted to reproduce a different setting? What if a friend wanted to reproduce the same level of lighting? You could say, "Turn the knob about a fifth of the way clockwise," or "Turn the knob until the arrow is at about two o'clock," but your friend's reproduction would only approximate your setting. What if your friend then passed the information on to another friend, who in turn passed it on again? Each time the information was handed on, the chances of its remaining accurate would decrease.

That's an example of information stored in "analog" form. The position of the dimmer's knob provides an analogy to the bulb's lighting level. If the knob is turned halfway, you presumably have about half the total wattage. When you measure or describe how far the knob is turned, you're actually storing information about the analogy (the knob's position) rather than about the lighting level. Analog information can be gathered, stored, and reproduced, but it tends to be imprecise—and you run the risk of its becoming less precise each time it's transferred.

Now let's look at an entirely different way of describing the room's light settings, a digital rather than an analog method of storing and transmitting information. Any kind of information can be converted into numbers made up of only 0s and 1s—binary numbers. Once the information has been converted into 0s and 1s, it can be fed to and stored in computers as long strings of bits. Those numbers are all that's meant by the term "digital information."

Let's say that instead of a single 250-watt bulb you have eight bulbs, each with a wattage double the one preceding it, from 1 through 128 watts. Each of these eight bulbs is hooked up to its own switch, with the lowest-watt bulb on the right. Such an arrangement can be diagrammed like this:

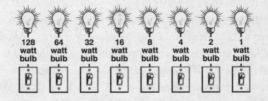

By turning these switches on and off, you can adjust the lighting level in 1-watt increments from 0 watt (all switches off) through 255 watts (all switches on). This gives you 256 precise possibilities. If you want 1 watt of light, you turn on only the rightmost switch, which turns on the 1-watt bulb. If you want 2 watts of light, you turn on only the 2-watt bulb. If you want 3 watts of light, you turn on both the 1-watt and 2-watt bulbs because 1 plus 2 equals the 3 watts you want. If you want 4 watts of light, you turn on the 4-watt bulb. If you want 5 watts, you turn on just the 4-watt and 1-watt bulbs. If you want 250 watts of light, you turn on all but the 4-watt and 1-watt bulbs.

- ▶

If you've decided that the ideal illumination level for dining is 137 watts of light, you turn on the 128-, 8-, and 1-watt bulbs, like this:

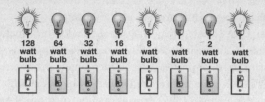

This system makes it easy to record an exact lighting level to use later or to communicate a lighting level to other people who have the same light switch setup. Because the way we record binary information is universal—low number to the right, high number to the left, always doubling—you don't have to write down the wattage values of the bulbs. You simply record the pattern of switches: on, off, off, off, on, off, off, on. With that information, a friend can faithfully reproduce the 137 watts of light ideal. As long as everyone involved double-checks the accuracy of what he does, the message can be passed through a million hands and at the end every person will have the same information and be able to achieve exactly 137 watts of light.

To shorten the notation further, you can record each "off" as 0 and each "on" as 1. This means that instead of writing down "on, off, off, off, on, off, off, on," meaning turn on (going from right to left) the first, the fourth, and the eighth of the eight bulbs and leave the others off, you can write the same information as 1, 0, 0, 0, 1, 0, 0, 1, or 10001001, a binary number. In this case it's 137. You call your friend and say: "I've got the perfect lighting level! It's 10001001. Try it." Your friend gets it exactly right by flipping a switch on for each 1 and off for each 0.

This may seem like a complicated way to describe the brightness of a light source, but it illustrates binary expression, the basis of all modern computers.

Binary expression made it possible to take advantage of electric circuits to build calculators. During World War II a group of mathematicians led by J. Presper Eckert and John Mauchly at the University of Pennsylvania's Moore School of Electrical Engineering began developing an electronic computational machine, the Electronic Numerical

Integrator And Calculator—ENIAC. Its purpose was to speed up the calculations for artillery-aiming tables. ENIAC was more like an electronic calculator than a computer, but instead of representing a binary number with on and off settings on wheels the way a mechanical calculator did, it used vacuum tube "switches."

Soldiers assigned by the army to the huge machine wheeled around squeaking grocery carts filled with vacuum tubes. When a tube burned out, ENIAC was shut down and the race to locate and replace the burned-out tube began. One explanation—it might be apocryphal—for why the tubes had to be replaced so often was that their heat and light attracted moths, which would fly into the huge machine and cause short circuits. If this is true, it's a source of the term "bugs," which is what we call the little glitches that can plague computer hardware and software.

When all of the tubes were working, a staff of engineers could set ENIAC up to solve a problem by laboriously plugging in 6,000 cables by hand. To make it perform another function, the staff had to reconfigure the cabling—every time. John von Neumann, a brilliant Hungarian-born American known for the development of game theory and his contributions to nuclear weaponry, among other things, is credited with the leading role in figuring out a way around this problem. He created the paradigm that all digital computers still follow. The "von Neumann architecture," as it is known today, is based on the principles he articulated in 1945—including the principle that a computer could avoid cabling changes by storing instructions in its memory. As soon as this idea was put into practice, the modern computer was born.

Today the brains of most computers are descendants of the microprocessor Paul Allen and I were so knocked out by in the 1970s, and personal computers are often rated according to how many bits of information (one switch in the lighting example) their microprocessors can process at a time, or how many bytes (a cluster of eight bits) of memory or disk-based storage they have. ENIAC weighed 30 tons and filled a 30-by-50–foot room. Inside, the computational pulses raced among 1,500 electromechanical relays and flowed through more than 18,000 vacuum tubes. Switching ENIAC on consumed 150,000 watts of energy. But ENIAC stored only the equivalent of about 80 characters of information.

1946: A view inside a part of the ENIAC computer

By the early 1960s, more than a decade after the discovery at Bell Labs that a tiny sliver of silicon could do the same job as a vacuum tube, transistors had supplanted vacuum tubes in consumer electronics. Like vacuum tubes, transistors act as electrical switches, but they require significantly less power to operate and as a result generate much less heat. And they require less space. Multiple transistor circuits can be combined onto a single chip, creating an integrated circuit. The computer chips we use today are integrated circuits containing the equivalent of millions of transistors packed onto less than a square inch of silicon.

In a 1977 *Scientific American* article, Bob Noyce, one of the founders of Intel, compared the $300 microprocessor to ENIAC, the moth-infested mastodon from the dawn of the computer age. The tiny micro-processor was not only more powerful, but as Noyce noted: "It is twenty times faster, has a larger memory, is thousands of times more reliable, consumes the power of a lightbulb rather than that of a locomotive, occupies 1/30,000 the volume and costs 1/10,000 as much. It is available by mail order or at your local hobby shop."

Of course, the 1977 microprocessor seems like a toy now. And in fact, many inexpensive toys contain computer chips more powerful than the 1970s chips that started the microcomputer revolution. But all of today's computers, whatever their size or power, manipulate information stored as binary numbers.

Binary numbers are used to store text in a personal computer, music on a compact disc, and money in a bank's network of cash machines. Before information can go into a computer, it has to be converted into binary numbers. You can imagine each device throwing switches in response to the binary numbers, controlling the flow of electrons. But the switches involved, which are usually made of silicon, are extremely small and can be thrown by applying electrical charges extraordinarily quickly—to produce text on the screen of a personal computer, music from a CD player, or the instructions to a cash machine to dispense currency.

The light switch wattage example demonstrated how any number can be represented in binary code. Here's how text can be expressed as binary numbers. By convention, the number 65 represents a capital A, the number 66 represents a capital B, and so on. Lowercase letters begin with number 97. On a computer the capital letter A, 65, becomes 01000001. The capital B, 66, becomes 01000010. A space break is represented by the number 32, or 00100000. So the sentence "Socrates is a man" becomes this 136-digit string of 1s and 0s:

01010011 01101111 01100011 01110010 01100001 01110100
01100101 01110011 00100000 01101001 01110011 00100000
01100001 00100000 01101101 01100001 01101110

To understand how other kinds of information are digitized, let's consider another example of analog information. A vinyl record is an analog representation of sound vibrations. It stores audio information in microscopic squiggles that line the record's long, spiral groove. If the music has a loud passage, the squiggles are cut more deeply into the groove, and if there is a high note, the squiggles are packed together more tightly. The groove's squiggles are analogs of the original vibrations—sound waves captured by a microphone. When a turntable's needle travels down the groove, it vibrates in resonation with the tiny squiggles. This vibration, still an analog representation of the original sound, is amplified and sent to loudspeakers as music.

Like any analog device for storing information, a vinyl record has drawbacks. Dust, fingerprints, or scratches on the record's surface can make the needle vibrate inappropriately and create clicks or other noises. If the record isn't turning at exactly the right speed, the pitch of the music won't be accurate. Each time a record is played, the needle wears away some of the subtleties of the squiggles in the groove and the quality of the reproduction of the music deteriorates. If you record a song from a vinyl record onto a cassette tape, any of the record's imperfections will be permanently transferred to the tape, and new imperfections will be added because conventional tape machines are themselves analog devices. The information loses quality with each generation of rerecording or retransmission.

On a compact disc, music is stored as a series of binary numbers, each bit (or switch) of which is represented by a microscopic pit on the surface of the disc. First-generation CDs, popular since the mid-1980s, have billions of pits. The reflected laser light inside the CD player—a digital device—reads each of the pits to determine whether it's switched to the 0 or the 1 position and then reassembles that information back into the original music by generating corresponding electrical signals that are converted by the speakers into sound waves. Each time the disc is played, the sounds are exactly the same.

It's convenient to be able to convert every kind of information into digital representations, but the number of bits can build up quite quickly. If there are too many bits of information, they can overflow the computer's memory or take a long time to transmit between computers. This is why a computer's capacity to compress digital data, store or transmit it, and then expand it back into its original form is so useful and will become even more so.

Quickly, here's how the computer accomplishes these feats. It goes back to Claude Shannon, the mathematician who in the 1930s recognized that information could be expressed in binary form. During World War II Shannon began developing a mathematical description of information and founded a field that later became known as information theory. He defined information as the reduction of uncertainty. By this definition, if you already know it's Saturday and somebody tells you it's Saturday, you haven't been given any information. On the other hand, if you're not sure of the day and somebody tells you it's Saturday, you've been given information—your uncertainty has been reduced.

Shannon's work on information theory eventually led to other breakthroughs. One was effective data compression, vital to both computing and communications. On the face of it, what Shannon said that led to data compression is obvious: Those parts of data that don't provide unique information are redundant and can be eliminated. Headline writers leave out nonessential words, as do people paying by the word to send a telegraph message or place a classified ad. One example of redundancy Shannon cited was the letter u, redundant in English whenever it follows the letter q. You know that a u will follow each q, so the u can be left out of a message without affecting its meaning.

In the half century since Shannon invented information theory and devised one of the first compression schemes, engineers have come up with brilliant ways of boiling redundancy out of information. It's not uncommon for the size of a text file to be cut roughly in half through compression. This lets it be transmitted twice as fast.

Shannon's principles have been applied to the compression of sound and picture data too. It's not unheard-of for a digital image to compress to a mere 5 percent of its original size after redundant information has been boiled out. If 12 pixels (the tiny picture elements, or dots, on a screen) in a row are the same color, it takes many fewer bits of information to describe the color once and indicate that it should be repeated a dozen times than it does to describe the color twelve times. Video typically contains a great deal of redundant information. Compression is achieved by storing information about how colors change—or don't change—from frame to frame, while storing information about the colors themselves only once every few frames.

The Internet makes use of compression, notably to transmit graphics, audio, and video on the World Wide Web, but compression alone won't meet the exploding need for communications capacity. We need to move ever-increasing numbers of bits from place to place. Bits travel through the air, through copper wires, and through fiber-optic cable ("fiber" for short). Fiber is cable made of glass or plastic so smooth and pure that, if you looked through a wall of it 70 miles thick, you'd be able to see a candle burning on the other side. Binary signals in the form of modulated light carry for long distances through these optic fibers. A signal doesn't move any faster through fiber-optic cable than it does through copper wire; both are governed by the speed of light. The enormous advantage fiber-optic cable offers over wire is its bandwidth. Bandwidth is a measure of the number of bits that can be moved through a circuit in a second. This really is like a highway. An eight-lane interstate has room for more vehicles than a narrow dirt road does. The more lanes available, the greater the bandwidth. Cables with limited bandwidth, used for text or voice transmissions, are called narrowband circuits. Cables with more capacity, which carry images and limited animation, have "midband capacity." Those with a high bandwidth, which can carry multiple video and audio signals, are said to have "broadband capacity."

Before a broadband interactive network is possible, fiber-optic cable must be brought into more neighborhoods and the performance and capacity of chips must continue to improve so that compression gets better and less expensive. Laying fiber will remain relatively expensive, but chips will get better and cheaper all the time.

In 1965 Gordon Moore, who later cofounded Intel with Bob Noyce, predicted that the capacity of a computer chip would double every year. Moore had looked at the price/performance ratio of computer chips—the amount of performance available per dollar—over the previous three years and simply projected it forward. Moore himself didn't believe that this rate of improvement would last long. But ten years later, his forecast proved out, and Moore then predicted that chip capacity would double every two years. To this day Moore's predictions have held up, and engineers now call the average rate of capacity increase—a doubling about every eighteen months—Moore's Law.

No experience in our everyday life prepares us for the implications of a number that doubles a great number of times—exponential increases. This fable explains it pretty well.

King Shirham of India was so pleased when one of his ministers invented the game of chess that he asked the man to name any reward.

"Your Majesty," said the minister, "I ask that you give me one grain of wheat for the first square of the chessboard, two grains for the second square, four grains for the third, and so on, doubling the number of grains each time until all sixty-four squares are accounted for." The king was moved by the modesty of the request and called for a bag of wheat.

The king asked that the promised grains be counted out onto the chessboard. On the first square of the first row was placed one small grain. On the second square were two specks of wheat. On the third square there were four—then on successive squares 8, 16, 32, 64, 128. By square eight at the end of the first row, King Shirham's supply master had counted out a total of 255 grains.

The king probably registered no concern. Maybe a little more wheat was on the board than he had expected, but nothing surprising had happened. If we assume that it took one second to count each grain, the counting so far had taken only about four minutes. If one row's grain could be counted out in four minutes, try to guess how long it would take to count out the wheat for all sixty-four squares of the board. Four hours? Four days? Four years?

By the time the second row was complete, the supply master had worked for about eighteen hours counting out just the sixteenth square's 65,535 grains. By the end of the third of the eight rows, it took ninety-seven days to count the 8.4 million grains for that twenty-fourth square. And there were still forty empty squares to go.

It's safe to assume that the king broke his promise to the minister.

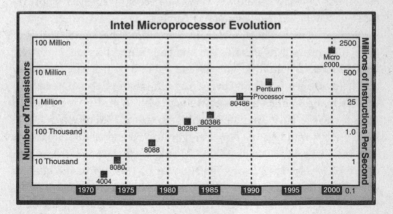

Intel microprocessors approximately doubling in transistor count every eighteen months, in accordance with Moore's Law

Completing the board would have required 18,446,744,073,709,551,615 grains of wheat and 584 billion years of counting. Current estimates of the age of the earth are around 4.5 billion years. According to most versions of the legend, King Shirham realized at some point in the counting that he had been tricked and had his clever minister beheaded.

Exponential growth, even when it's explained, seems like a trick.

Moore's Law is likely to hold for another twenty years. If it does, a computation that now takes a day will be more than 10,000 times faster and thus take fewer than ten seconds.

Laboratories are already operating "ballistic" transistors that have switching times on the order of a femtosecond. That's 1/1,000,000,000,000,000 of a second, which is about 10 million times faster than the switching time for transistors in one of today's microprocessors. The trick is to reduce the size of the chip circuitry and the current flow so that moving electrons don't bump into anything, including each other. The next stage is the "single-electron transistor," in which a single bit of information is represented by a lone electron. This will be the ultimate in low-power computing, at least according to our current understanding of physics. In order to make use of the incredible speed advantages to be had at the molecular level, computers will have to be very small, even microscopic. We already understand the science that would allow us to build these superfast computers. What we need is an engineering breakthrough, and these are often quick in coming.

Storing all those bits shouldn't be a problem either. In the spring of 1983 IBM released its PC/XT, the company's first personal computer with an internal hard disk. The disk served as a built-in storage device and held 10 megabytes, or "megs," of information, about 10 million characters, or 80 million bits. Customers who wanted to add these 10 megs to their original computers could, for a price. IBM offered a $3,000 kit, complete with separate power supply, for expanding the computer's storage. That's $300 per megabyte. Thanks to the exponential growth described by Moore's Law, by the summer of 1996 personal computer hard drives that could hold 1.6 gigabytes—1.6 billion characters of information—were priced at $225. That's $0.14 per megabyte! And we're expecting an exotic improvement called a holographic memory, which can hold terabytes of characters in less than a cubic inch of volume. With such a capability, a holographic memory the size of your fist could hold the contents of the Library of Congress.

As communications technology goes digital, it stands to benefit from the same exponential improvements that have made today's $2,000 laptop computer more powerful than a $10 million IBM mainframe computer of twenty years ago.

At some point a single wire running into each home will be able to send and receive all of a household's digital data. The wire might be fiber, which is what long-distance telephone calls are carried on now; or coaxial cable, which currently brings us cable television signals; or the simple "twisted-pair" wire that connects telephones in homes to the local phone system. It may even be a wireless connection. If the bits coming into the house are interpreted as voice calls, the phone will ring. If they're video images, they'll show up on the television set or a PC. If they're news, they'll arrive as text and pictures on a screen.

That single connection to the network will certainly carry much more than phone calls, movies, and news. But we can no more imagine what the broadband information highway will carry in twenty-five years than a Stone Age man using a crude knife could have envisioned Ghiberti's Baptistery doors in Florence. Only as the Internet evolves will all of the possibilities be understood.

3

LESSONS FROM THE COMPUTER INDUSTRY

Success is a lousy teacher. It seduces smart people into thinking they can't lose. And it's an unreliable guide to the future. What seems to be the perfect business plan or the latest technology today may soon be as out-of-date as the eight-track tape player, the vacuum-tube television, or the mainframe computer. I've watched it happen. History is a good teacher, though, and observing many companies over a long period of time can teach us principles that will help us with strategies for the years ahead.

Companies investing in interactive networks will try to avoid repeating the mistakes made in the computer industry over the past twenty years. I think most of these mistakes can be understood by looking at a few critical factors: negative and positive spirals, the need to initiate rather than follow trends, the importance of software as opposed to hardware, and the role of compatibility and the positive feedback it can generate.

You can't count on conventional wisdom, which makes sense only in conventional industries. For the last three decades the behavior of the computer hardware and software industries has definitely been unconventional. Big established companies that had hundreds of millions of dollars in sales and lots of satisfied customers have disappeared in a short time. New companies, such as Apple, Compaq, Lotus, Oracle, Sun, and Microsoft, appeared to go from nothing to a billion dollars of revenue in a flash. These successes were driven in part by what I call a "positive spiral."

When you have a hot product, investors pay attention to you and are willing to put their money into your company. Smart kids think, Hey, everybody's talking about this company. I'd like to work there. When one smart person comes into a company, pretty soon another does because talented people like to work with each other. This creates a sense of excitement. Potential partners and customers pay more attention, and the spiral continues, making the next success easier.

Of course, companies can get caught in a negative spiral too. A company in a positive spiral has an air of destiny. One in a negative spiral operates in an atmosphere of doom. If a company starts to lose market share or delivers a bad product, the talk turns to "Why do you work there?" "Why would you invest in that company?" "I don't think you should buy from them." The press and the analysts smell blood and start telling inside stories about who's quarreling and who's responsible for mismanagement. Customers begin to question whether they should continue to buy the company's products. Within the sick company everything gets questioned, including the things the company is doing right. Even a fine strategy can get dismissed with the argument "You're just defending the old way," and that can lead to more mistakes. Then down the company spirals. Leaders such as Lee Iacocca, who was able to reverse a negative spiral, deserve a lot of credit.

While I was growing up, the hot computer company was Digital Equipment Corporation. For twenty years DEC's positive spiral seemed unstoppable. Ken Olsen, the company's founder, was a legendary hardware designer and a hero of mine, a distant god. In 1960 he had created the minicomputer industry by offering the first "small" computers. The earliest was the PDP-1, the ancestor of my high school's PDP-8. Instead of paying the millions IBM wanted for its "Big Iron," a buyer could get one of Olsen's PDP-1s for $120,000. It wasn't nearly as powerful as the big machines, but it could be used for a wide variety of applications that didn't need the computing power of a mainframe. DEC grew to be a $6.7 billion company in eight years by offering a wide range of computers in different sizes.

Two decades later Olsen's vision wavered. He couldn't see the future in small desktop computers. Eventually he was forced out of DEC, and part of Olsen's legend now is that he is the man famous for repeatedly, and publicly, dismissing the personal computer as a passing fad. I'm sobered by stories like Olsen's. He was brilliant at seeing new ways of doing things, and then—after years of being an innovator—he missed a big bend in the road.

Another visionary who faltered was An Wang, the Chinese immigrant to the U.S. who built Wang Laboratories into the dominant supplier of electronic calculators in the 1960s. In the 1970s Wang ignored the advice of everyone around him and left the calculator market just before the arrival of low-cost competition that would have ruined him. It was a brilliant move. Wang reinvented his company to be the leading supplier of word processing machines. During the 1970s, in offices around the world, Wang word processing terminals began to replace typewriters. The machines contained a microprocessor, but they weren't true personal computers because they were designed to do only one thing—handle text.

The kind of insight that had led Wang to abandon calculators could have led his company to success in personal computer software in the 1980s, but he didn't spot the next industry turn. Even though he developed great software, it was tied proprietarily to his word processors. Once general-purpose personal computers appeared, Wang's software and his single-purpose machine were doomed. PCs could run a variety of word processing software applications, such as WordStar, Word-Perfect, and MultiMate (which imitated Wang software), plus other kinds of applications. If Wang had recognized the importance of widely compatible software applications and general-purpose computers, there might not be a Microsoft today. I might be a mathematician or an attorney somewhere, and my adolescent foray into personal computing might be little more than a distant personal memory.

IBM is another major company that misunderstood the technological and market forces that were bringing about the PC revolution. For decades the company's leader was Thomas J. Watson, a hard-driving former cash-register salesman. Watson wasn't the founder of IBM, but it was thanks to his aggressive management that by the early 1930s IBM dominated the market for accounting machines.

IBM had begun working on computers in the mid-1950s. It was one of many companies vying for leadership in the new field. Until 1964 each computer model, even from the same manufacturer, had had a unique design and required its own operating system and application software. An operating system (sometimes called a disk-operating system, or just DOS) is the fundamental software that tells a computer system's components how to work together and performs other broad functions. It's a platform on which all the software applications—such as accounting or payroll or word processing or electronic mail programs—are built. Without an operating system, a computer is useless.

Computers at different price levels had different designs. Some models were dedicated to scientific study, others to commerce. As I would discover in the 1970s, when I wrote versions of BASIC for various personal computers, a significant amount of work was required to enable software written to run on one computer model to run on another. This was true even if the software was written in a standard language such as COBOL or FORTRAN.

Under the direction of "Young Tom," as Watson's son and successor was known, the company gambled $5 billion in the mid-1960s on the novel notion of scalable architecture—all the computers in the System/360 family, no matter what size, would respond to the same set of instructions. Models built with different technology, from the slowest to the fastest, from the smallest to the largest, would run the same operating system. Customers would be able to move their applications and their hardware peripherals—accessories such as storage disks, tape drives, and printers—freely from one model to the next. IBM's notion of scalable architecture completely reshaped the industry.

System/360 was a runaway success and made IBM the powerhouse in mainframe computers for the next thirty years. Customers made big investments in the 360, confident that their commitment to software and training for it wouldn't be wasted. If they needed to move to a larger computer, they could get a bigger IBM that ran the same operating system and application software and shared the same architecture. In 1977 DEC introduced its own scalable-architecture platform, the VAX. The VAX family of computers ultimately ranged from desktop systems to mainframe-size machine clusters and did for DEC what the System/360 family did for IBM. DEC became the overwhelming leader in the minicomputer market.

The scalable architecture of the IBM System/360 and its successor, the System/370, drove many of IBM's competitors out of business and scared away potential newcomers—at least for a while. Then in 1970 Gene Amdahl, who had been a senior engineer at IBM, founded a competing company. Amdahl had a novel business plan. His Amdahl company would build computers fully compatible with the IBM 360 software. Amdahl delivered hardware that not only ran the same operating systems and applications as IBM's 360 machines but, because it took advantage of new technology, also outperformed IBM's comparably priced systems. Soon Control Data, Hitachi, and Itel (not Intel) all offered mainframes that were "plug-compatible" with IBM's machines. By the mid-1970s the importance of 360 compatibility was becoming obvious. The only mainframe companies doing well were those whose hardware could run IBM's operating systems.

Before the 360 and its "clones," computers were intentionally designed to be incompatible with those from other companies because the manufacturer's goal was to make it discouragingly difficult and expensive for existing customers to switch to a different brand. Once a customer committed to a machine, he or she was stuck with offerings from the computer's manufacturer because changing the software, while it could be done, was difficult and expensive. Amdahl and the other IBM-compatible companies ended that stranglehold on the customer. Now customers could—and did—choose systems that gave them a choice of hardware suppliers and the widest variety of software applications. Market-driven compatibility proved to be an important lesson for the mainframe and minicomputer industries, and later it was an important lesson for the personal computer industry. It promises to be an important lesson for the Internet industry as well.

While the computer industry was learning about the importance of compatibility, I was busy with lessons of a different kind. I had arrived at Harvard in the fall of 1973 and had been an economics major before switching to mathematics. I found mathematics fascinating, but the difficulty of making a new discovery in mathematics and the relatively small impact such a discovery would have in the wider world made me wonder if that was how I wanted to spend my life. I considered psychology, and I kept the possibility of studying law in the back of my mind. In any case, I wasn't all that worried about it. I was having fun.

In college a lot of posturing goes on, and looking as if you don't need to study, as if you can slack off, can be considered a great way to establish your coolness. During my freshman year I instituted a deliberate policy of skipping most classes and then studying feverishly at the end of the term. It became a game—a not uncommon one—to see how high a grade I could pull while investing the least time possible.

I filled a lot of my free time playing poker, which had its own attraction. In poker, a player collects different pieces of information—who's betting boldly, what cards are showing, what this guy's pattern of betting and bluffing is—and then crunches all that data together to devise a plan for his own hand. I got pretty good at this kind of information processing. I played about three times a week. Games usually included half a dozen other students and lasted well into the night—or the next day. It wasn't unusual for me to win or lose $500 or $600 a night. More than once I found myself thinking, "Oh God, it's eight o'clock in the morning and I've been sitting at this poker table for twelve hours and I was supposed to do my homework. Why did I stay in the game? I've lost $400 and I feel awful. But if I play one more hand, maybe I'll have lost only $300." It wasn't always fun.

The poker strategizing experience would prove helpful when I got into business, though. The other game I played at college, the postponing one, wouldn't serve me well at all. I didn't know that back then. In fact, I was encouraged in my dilatory practices because they were shared by Steve Ballmer, a math major I met freshman year, when we both lived in Currier House. Steve and I led very different lives, but we were both trying to pare down to a minimum the course and studying time we needed to get top grades. Steve is a man of boundless energy, effortlessly gregarious, and his extracurricular activities took up a lot of his time. By his sophomore year he was a manager of the football team, the advertising manager for the *Harvard Crimson*, and president of a literary magazine. He also belonged to a social club, the Harvard equivalent of a fraternity.

Steve and I would pay very little attention to our classes and then furiously inhale the key books just before an exam. Once we took a tough graduate-level economics course together—Economics 2010. The professor allowed you to bet your whole grade on the final if you chose. So Steve and I focused on other areas all semester and did absolutely nothing for the course until the week before the last exam. Then we studied like mad and ended up getting A's.

After Paul Allen and I started Microsoft, I found out that developing the habit of procrastination hadn't been the best preparation for running a company. Among Microsoft's first customers were companies in Japan so methodical that the minute we got behind schedule they would fly someone over to baby-sit us. They knew that their man couldn't really help, but he stayed in our office eighteen hours a day just to show us how much they cared. These guys were serious! They would ask, "Why did the schedule change? We need a reason. And we're going to change the thing that caused it to happen." I can still feel how painful being late on some of those projects got to be. We did mend our ways. We're still late with projects sometimes but a lot less often than we would have been if we hadn't had those intimidating baby-sitters.

In 1975 Microsoft had started out in Albuquerque, New Mexico, because that's where MITS, the little company whose Altair 8800 personal computer kit had been on the cover of *Popular Electronics*, was located. We worked with MITS because it was the first company to sell an inexpensive personal computer to the general public. In return for our software, MITS gave us royalties and office space the first year we were in Albuquerque.

But after it was acquired by another company, MITS stopped paying us. We had no income for a year and were basically broke. An arbitrator handling the dispute was four months late in his ruling, and we were hanging on, but only barely. I talked to my father about the legal case during that time, and his advice helped me stick it out. I could have asked Dad for a loan, but I never did. Our attorney was nice enough to wait for payment until we won the case. After that episode, Microsoft has been perpetually cash-flow positive. In fact, I developed a rule: We always have to have enough *cash* on hand to be able to run the company for at least a year even if no one pays us. The MITS experience, suddenly having no income, made me very conservative financially, a trait that persists to this day.

By 1977 Apple, Commodore, and Radio Shack had also entered the personal computer business. Microsoft provided BASIC for most of the early personal computers. The BASIC language was the crucial software ingredient at that time because users wrote their own applications in BASIC instead of buying packaged applications.

In the early days selling BASIC was one of my many jobs. For the first three years most of the other professionals at Microsoft focused solely on the technical work, and I did most of the sales, finance, and marketing as well as some of the code writing. I was barely out of my teens, and selling intimidated me. Microsoft's strategy was to get computer companies such as Radio Shack to license our software so that they could include it with the personal computers they sold and pay us a royalty.

Within a few months of leaving Harvard I found myself in Texas selling BASIC to Radio Shack for its TRS-80 personal computer. John Roach, then the company's vice president and now its CEO, asked me how much I wanted.

"Fifty thousand dollars," I said.

"Horseshit!" Roach replied.

I held my ground with Roach, arguing that software was a crucial part of what Radio Shack's customers would come to expect with their personal computers. Roach was a formidable guy, but he gave me my price.

Each time I succeeded in selling BASIC to a computer company, I got a little more confident. When it came time to sell to Texas Instruments, I decided that $100,000 was fair, but I was afraid they might balk at six figures. So I offered them a grocery store bargain. Only $99,000! It was a deal, and they bought it.

These prices may sound high, but they were really a steal. The fees we charged were far below what it would cost a company such as Radio

Shack to create its own programming language. In fact, the prices I offered were lower than the cost a company might naively fool itself into believing it could achieve on its own. Almost every computer manufacturer licensed BASIC from us.

One reason we wanted to sell to computer companies rather than consumers was software piracy. We wanted to get paid for our work, and when companies bundled our software with their computers, they included our royalty in the price. In the early years of selling Altair BASIC directly to consumers, our sales had been far lower than the widespread use of our software suggested they should be. I wrote a widely disseminated "Open Letter to Hobbyists" asking the early users of personal computers to stop stealing our software so that we could make money that would let us build more software. "Nothing would please me more than being able to hire ten programmers and deluge the hobby market with good software," I wrote. But my argument didn't convince many hobbyists to pay for our work. They seemed to like it, and they used it, but they seemed to prefer to "borrow" it from each other.

Fortunately, most users today understand that software is protected by copyright. Software piracy is still a major issue in trade relations because some countries still don't recognize—or don't enforce—the international standard for copyright laws. The United States urges other governments to do more to enforce copyright laws for books, movies, CDs, and software. This issue will only get more important as we move further into the global information economy.

Although we were successfully selling our software to U.S. hardware companies, by 1979 almost half of our business was coming from Japan, thanks to an extraordinary guy named Kazuhiko (Kay) Nishi. Kay telephoned me in 1978 and introduced himself in English. He had read about Microsoft and thought he should be doing business with us. As it happened, Kay and I had a lot in common. We were the same age, and he too was a college student on leave because of his passion for personal computers.

Kay and I met some months later at a computer conference in Anaheim, California, and he flew back with me to Albuquerque, where we signed a page-and-a-half contract that gave him exclusive distribution rights for Microsoft BASIC in East Asia. There were no attorneys involved, just Kay and I, kindred spirits. We did more than $150 million of business under that contract—more than ten times what we had expected.

Kay moved fluidly between the business cultures of Japan and the United States. He was flamboyant, which worked in our favor in Japan because it bolstered the impression among Japanese businessmen that we were whiz kids. When I was in Japan, we'd stay in the same hotel room and he'd be getting phone calls all night long, booking millions of dollars of business. It was phenomenal. One time there were no calls between three and five in the morning. When a call came in at five o'clock, Kay reached for the phone and said, "Business is a little slow tonight." It was quite a ride.

For the next eight years Kay seized every opportunity. Once in 1981 on a flight from Seattle to Tokyo, Kay found himself sitting next to Kazuo Inamori, president of the $650 million Kyocera Corporation. Kay, who ran his Japanese company, ASCII, confident of Microsoft's cooperation, successfully pitched Inamori on a new idea—a small laptop computer with simple software built in. Then Kay and I designed the machine, and Microsoft was still small enough that I could play a personal role in the software development for it. In the United States, our little machine was marketed by Radio Shack in 1983 as the Model 100 for as little as $799. It was also sold in Japan as the NEC PC-8200 and in Europe as the Olivetti M-10. Thanks to Kay's enthusiasm, it was the first popular laptop, a favorite of journalists for years.

Years later, in 1986, Kay decided he wanted to take ASCII in a direction different from the one I wanted for Microsoft, so Microsoft set up its own subsidiary in Japan. Kay's company has continued to be a very important publisher in the Japanese market. Kay is a close friend, and he's still flamboyant and committed to making personal computers universal tools.

The global nature of the PC market will also be a vital element in the development of the information highway. Collaboration among American, European, and Asian companies will be even more important for the personal computer than they have been in the past. Countries and companies that fail to make their work global won't be able to lead.

In January 1979 Microsoft moved from Albuquerque to a suburb of

Seattle, Washington. Paul and I had come home, bringing almost all of our dozen employees with us. Our main business was writing programming languages for the profusion of new machines that appeared as the personal computer industry took off. People were coming to us with all kinds of interesting projects that had the potential to turn into something big. Demand for Microsoft's services exceeded what we could supply.

I needed help running the business and turned to Steve Ballmer, my old Harvard Economics 2010 pal. After graduating Steve had worked as an associate product manager for Procter & Gamble, where his work had included a stint paying calls on small grocery stores. After a few years he'd decided to go to the Stanford Business School. When he got my call, he had finished only one year and wanted to complete his degree, but when I offered him part ownership of Microsoft, he became another student on indefinite leave. Shared ownership through the stock options Microsoft offered most of its employees has contributed more to our success than anyone would have predicted. Employee shares are a great motivator, and Microsoft employees have realized literally billions of dollars from their stock options over the years. The practice of granting employee stock options, which has been widely and enthusiastically adopted in the U.S., has supported a disproportionate number of start-up successes and will enable new startups to seize the opportunities coming up.

Within three weeks of Steve's arrival at Microsoft we had the first of our very few arguments. Microsoft employed about thirty people by this time, and Steve had concluded that we needed to add fifty more immediately.

"No way," I said. Many of our early customers had gone bankrupt, and after our MITS experience I still dreaded going bust in a boom time. I wanted Microsoft to be lean and hungry. But Steve wouldn't relent, so I did. "Just keep hiring smart people as fast as you can," I said, "and I'll tell you when you get ahead of what we can afford." I never had to put a halt to all that hiring because our income grew as fast as Steve could find great people.

My chief fear in the early years was that some other company would swoop in and take the market from us. Several small companies making either microprocessor chips or software had me particularly worried, but luckily for us none of them saw the software market quite the way we did.

There was also always the threat that one of the major computer manufacturers would take the software for their larger machines and scale it down to run on small microprocessor-based computers. IBM and DEC had libraries of powerful software. Fortunately for Microsoft again, the major players never focused on bringing their computer architecture and software over to the personal computer. The only close call came in 1979, when DEC offered the PDP-11 minicomputer architecture in a personal computer kit marketed by HeathKit. DEC didn't completely believe in personal computers, though, and didn't really push the product.

Microsoft's goal was to write and supply software for most personal computers without getting directly involved in making or selling computer hardware. We licensed our software at extremely low prices because we believed money could be made betting on volume. We'd adapt a programming language, such as our version of BASIC, to each machine that came on the market. We were very responsive to all the hardware manufacturers' requests. We didn't want to give anybody a reason to look elsewhere. We wanted choosing Microsoft software to be a no-brainer.

Our strategy worked. Virtually every personal computer manufacturer licensed a programming language from us. Even though the hardware architecture of two companies' computers might be different, the fact that computers from both companies ran Microsoft BASIC meant that their machines were somewhat compatible. That compatibility became an important part of what people purchased with their computers. Manufacturers frequently advertised that Microsoft programming languages, including BASIC, were available for their computers.

Along the way, Microsoft BASIC became an industry software standard.

Some technologies don't depend on widespread acceptance for their value. A wonderful nonstick frying pan is useful even if you're the only person who ever buys one. But for communications products and other products that involve collaboration, much of a product's value comes from its widespread deployment. Given a choice between a beautiful, handcrafted mailbox with an opening that would accommodate only one size of envelope, and an old carton that everyone routinely dropped all mail and messages for you into, you'd choose the one with broader access. You'd choose compatibility.

To promote compatibility, governments or committees sometimes set standards. These "de jure" standards have the force of law. Many of the most successful standards, however, are "de facto"—standards the market discovers. Most analog timepieces operate clockwise. English-language typewriter and computer keyboards use a layout in which the first six keys across the top row of letters, left to right, spell QWERTY. No law enforces these standards. They're widespread and they work, and most customers will stick with them unless something dramatically better comes along.

Because de facto standards get developed by the marketplace rather than by law, they tend to be chosen for the right reasons and to get replaced when something truly better shows up—the way the compact disc (CD) has replaced the vinyl record.

De facto standards often evolve in the marketplace through an economic mechanism similar to the positive spiral that drives successful businesses—success reinforces success.

A de facto standard emerges when, in a growing market, one way of doing something gets a slight advantage over competitive ways of doing that thing. It's most likely to happen with a high-technology product that can be made in great volume at a very small marginal cost and that derives some of its value from compatibility. A home video game system is a special-purpose computer equipped with a special-purpose operating system that forms a platform for the game's software. Compatibility is important because the greater the number of software applications—in this case, games—that are available for it, the more valuable the machine becomes to a consumer. And the more machines consumers buy, the more applications software developers create for it. A positive-feedback cycle sets in once the machine reaches a high level of popularity, and sales continue to grow.

One of the most famous demonstrations of the power of positive feedback was the videocassette recorder format battle of the late 1970s and early 1980s. The persistent myth has been that positive feedback alone caused the VHS format to win out over the Beta format even though Beta was technically better. Actually, early Beta tapes recorded for only an hour, not enough time for a whole movie or football game. Customers care more about a tape's capacity than about engineering specifications, so the VHS format, which allowed longer recordings, got off to a small lead over the Beta format used by Sony in its Betamax player. JVC, the developer of the VHS standard, allowed other VCR manufacturers to use the VHS standard for a very low royalty. As VHS-compatible players proliferated, video rental stores started to stock more VHS than Beta tapes. The owner of a VHS player was thus more likely than a Beta owner to find the movie she wanted at the video store, which made VHS fundamentally more useful to its owners, causing even more people to buy VHS players. This further motivated video stores to stock VHS tapes. Beta lost out as people chose VHS in the belief that it represented a durable standard. VHS was the beneficiary of a positive-feedback cycle. Success bred success.

While the duel between the Betamax and VHS formats was at its height, sales of prerecorded videocassettes to U.S. tape rental dealers were almost flat, just a few million copies a year. Once VHS emerged as the apparent standard, in about 1983, an acceptance threshold was crossed and the use of the machines as measured by tape sales turned abruptly upward. That year over 9.5 million tapes were sold, a more than 50 percent increase over the year before. In 1984 tape sales reached 22 million and, in successive years, 52 million and 84 million, reaching 110 million units in 1987, by which time renting movies had became one of the most popular forms of home entertainment and the VHS machine had become ubiquitous.

This is an example of how a quantitative change in the acceptance level of a new technology can lead to a qualitative change in the role the technology plays. Television provides another. In 1946, 10,000 television sets were sold in the United States, and only 16,000 were sold in the next year. But then a threshold was crossed. In 1948 the number was 190,000. In successive years it was 1 million units, followed by 4 million, 10 million, and steadily up, until 32 million were sold in 1955. As more television sets were sold, more programming was created, which enticed more people to buy television sets.

For the first few years after they were introduced, audio compact disc players and the discs themselves didn't sell well, in part because it was difficult to find music stores that carried many CD titles. Then seemingly overnight enough players had been sold, more titles were available, and an acceptance threshold had been crossed. More people bought players because more titles were available, and record companies made more titles available on CDs. Most music lovers preferred the new, high-quality sound and convenience of compact discs, and CDs became the de facto standard, driving LPs out of the record stores.

One of the most important lessons the computer industry learned early on is that a great deal of a general-purpose computer's value to its user depends on the quality and variety of the application software available for it. All of us in the industry learned that lesson—some happily, some unhappily.

In the summer of 1980 two emissaries from IBM came to Microsoft to discuss a personal computer they said IBM might or might not build.

At the time IBM's position in the realm of big hardware was unchallenged. It had a more than 80 percent share of the large computer market. Used to selling big, expensive machines to big customers, IBM had had only modest success with smaller computers. Its management suspected that IBM, even though it had 340,000 employees, would need the assistance of outsiders if it was going to sell little, inexpensive machines to individuals as well as companies anytime soon.

IBM wanted to bring its personal computer to market in less than a year. In order to meet this schedule, it had been forced to abandon its traditional course of creating all the hardware and software itself. IBM had decided to build its PC mainly from off-the-shelf components available to anybody. This created an essentially open architecture, easy to copy.

Although it generally built the microprocessors used in its products, IBM decided to buy microprocessors for its PCs from Intel. Most important for Microsoft, IBM decided to license the operating system from us rather than create the software itself.

*1981: The IBM
Personal Computer*

Working with the IBM design team, we encouraged IBM to build one of the first personal computers to use a 16-bit microprocessor chip, the Intel 8088. We could see that the move from 8 to 16 bits would take the personal computer from hobbyist toy to high-volume business tool. The 16-bit generation of computers could support up to one full megabyte of memory—256 times as much as an 8-bit computer. At first this would be just a theoretical advantage because IBM intended to offer only 16 kilobytes (16K) of memory initially—1/64 of the total memory possible. The benefit of going to 16 bits was further diluted by IBM's decision to save money by using a chip that employed only 8-bit connections to the rest of the computer—consequently, the chip could think much faster than it could communicate.

We could see that, with its reputation and its decision to employ an open design that other companies could copy, IBM had a real chance to create a new, broad standard in personal computing. We wanted to be a part of it, so we took on the operating system challenge. We bought some early work from another Seattle company and hired its top engineer, Tim Paterson. With lots of modifications, the system became the

Microsoft Disk Operating System, or MS-DOS. Tim became, in effect, the father of MS-DOS.

IBM, our first licensee, called the system PC-DOS—"PC" for "Personal Computer." The IBM Personal Computer hit the market in August 1981 and was an immediate triumph. The company did a good job of marketing the machine and popularized the term "PC." The project had been conceived by Bill Lowe and shepherded to completion by Don Estridge. It's a tribute to the quality of the IBM people involved in the project that they were able to take their personal computer from idea to market in less than a year.

Few people remember this now, but the original IBM PC actually shipped with a choice of three operating systems—our PC-DOS, Digital Research's CP/M-86, and the UCSD Pascal P-system. We knew that only one of the three could succeed and become the standard. We wanted the same kinds of forces that were putting VHS cassettes into every video store to push MS-DOS to become the standard. We saw three ways to get MS-DOS out in front. First was to make MS-DOS the best product. Second was to help other software companies write MS-DOS–based applications software. Third was to ensure that MS-DOS would be inexpensive to license.

We gave IBM a fabulous deal—a one-time fee of about $80,000 that granted the company the royalty-free right to use Microsoft's operating system forever. In other words, we practically gave the software to IBM. Giving software away to create strategic value has since become a well-established marketing technique in the industry, but it was uncommon at the time. The deal gave IBM an incentive to push MS-DOS and to sell it inexpensively. Our strategy worked. IBM sold the UCSD Pascal P-System for about $450, CP/M-86 for about $175, and MS-DOS for about $60.

Our goal was not to make money directly from IBM but to profit from licensing MS-DOS to computer companies who wanted to offer machines more or less compatible with the IBM PC. IBM could use our software for free, but it didn't have an exclusive license or control of future enhancements. This put Microsoft in the business of licensing operating system software to the personal computer industry. Eventually IBM abandoned the UCSD Pascal P-system and CP/M-86 operating systems.

Consumers bought the IBM PC with confidence, and in 1982 third party software developers began turning out applications to run on it. Each new customer and each new application added to the IBM PC's strength as a potential de facto standard for the industry. Soon most of the new and best software, such as the revolutionary spreadsheet Lotus 1-2-3, was being written for it. The original inventors of the electronic spreadsheet, Dan Bricklin and Bob Frankston, deserve immense credit for their VisiCalc product. But 1-2-3 made it obsolete by embracing what VisiCalc did well and extending its functionality in compelling ways that took advantage of the IBM PC's strengths. Mitch Kapor, who created 1-2-3 with Jonathan Sachs, is a fascinating person whose eclectic background—he had been a disc jockey and a transcendental meditation instructor—is typical of the best software designers.

A positive-feedback cycle began to drive the PC market. Once it got going, thousands of software applications appeared, and untold numbers of companies began making add-in, or "accessory," cards that extended the hardware capabilities of the PC. The availability of software and of hardware add-ons helped PCs sell at a far greater rate than IBM had anticipated. The positive-feedback cycle spun out billions of dollars for IBM. For a few years more than half of all personal computers used in business were IBMs, and most of the rest were compatible with IBM machines.

The IBM standard became the platform everybody imitated. IBM's prestige made the PC the seed corn for the broad standard. IBM's early timing and its use of a 16-bit processor were critical factors too. Both timing and marketing are key to acceptance with technology products. The PC happened to be a good machine, but it wasn't a given that IBM would set the standard. Under different circumstances, another company might have achieved critical mass by promoting the creation of enough good applications and getting enough machines sold.

IBM's early business decisions, which grew out of its rush to get the PC to market, made it easy for other companies to build compatible machines. The architecture was for sale. The microprocessor chips from Intel and Microsoft's operating system were available to any startup. This openness was a powerful incentive for component builders, software developers, and everybody else in the business.

---------------------------▶

Within three years almost all of the competing standards for personal computers had disappeared. The only exceptions were Apple's Apple II and Macintosh. Hewlett-Packard, DEC, Texas Instruments, and Xerox, despite their good technologies, reputations, and customer bases, failed in the personal computer market in the early 1980s because their machines weren't compatible with IBM's and didn't offer significant enough improvements over the IBM architecture to encourage a shift. Many startups, such as Eagle and Northstar, thought people would buy their hardware because it offered something different and slightly better than the IBM PC. They were wrong. The IBM PC had become the hardware standard, and all of the startups either changed over to building compatible hardware or failed. By the mid-1980s there were dozens of IBM-compatible PCs. Although buyers of a PC might not have articulated it this way, what they were looking for was the hardware that ran the most software, and they wanted the same system the people they knew and worked with had.

Some commentators like to conclude that IBM made a mistake working with Intel and Microsoft to create its PC. They argue that IBM should have kept the PC architecture proprietary and that Intel and Microsoft somehow got the better of IBM. But the commentators are missing the point. IBM became the central force in the PC industry precisely because it was able to harness an incredible amount of innovative talent and entrepreneurial energy and use it to promote its open architecture. IBM set the standards.

In the mainframe business, IBM was king of the hill, and competitors found it hard to match the IBM sales force and IBM's investment in research and development. If a competitor tried climbing the hill, IBM could focus its assets to make the ascent nearly impossible. But in the volatile world of the personal computer, IBM's position was more like that of the leading runner in a marathon. As long as the leader keeps running as fast as or faster than the others, he stays in the lead and competitors will have to keep trying to catch up. If the leader slacks off or stops pushing himself, though, the rest of the field will pass him by. IBM could keep its lead only by running hard.

By 1983 I thought Microsoft's next step should be to develop a graphical operating system. I didn't believe we'd be able to retain our position at the forefront of the software industry if we stuck with MS-DOS because MS-DOS was character-based. A user had to type in often-obscure commands, which then appeared on the screen. MS-DOS didn't provide pictures and other graphics to help users communicate with applications. I believed that interfaces—the "personalities" of programs—would be graphical in the future and that it was essential for Microsoft to move beyond MS-DOS and set a new standard in which pictures and better-looking fonts (typefaces) would be parts of an easier-to-use interface. If we were to realize our vision of widespread personal computer use, PCs had to be made easier to use—not only to help existing customers but also to attract new customers who might not take the time to learn to work with a complicated interface.

To see the huge difference between a character-based computer program and a graphical one, imagine playing a board game such as chess, checkers, Go, or Monopoly on a computer screen. With a character-based program, you type in your moves using characters. You write "Move the piece on square 11 to square 19," or something slightly more cryptic like "Pawn to QB3." But in a graphical computer program, you

1984: Character-based user interface in an early PC word processor

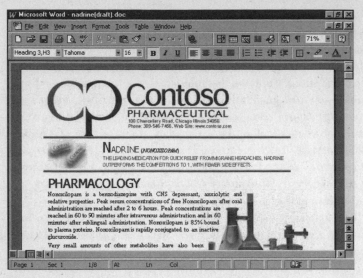

1996: Graphical user interface in a modern word processor

see the board game on your screen. You move pieces by pointing at them and actually dragging them to their new locations.

Researchers at Xerox's now-famous Palo Alto Research Center (PARC) in California had explored new paradigms for human-computer interaction. Their research had shown that it was easier to give a computer instructions if you could see pictures and point at things on the screen. They used a device called a "mouse," which could be rolled on a tabletop to move a pointer around on the screen. Xerox did a poor job of taking commercial advantage of this groundbreaking idea because its machines were expensive and didn't use standard microprocessors. Translating great research into products that sell is still a big problem for many companies.

In 1983 Microsoft announced that we planned to bring graphical computing to the IBM PC with a product called Windows. Our goal was to create software that would extend MS-DOS and let people use a mouse. We'd use graphical images on the computer screen and make a number of "windows" available on the screen, each of which could run a different computer program. At that time two of the personal computers on the market had graphical capabilities: the Xerox Star and the Apple Lisa. Both were expensive, had limited capability, and were built on proprietary hardware architectures. Other hardware companies couldn't license the Xerox or Apple operating systems to build compatible computers, and neither platform attracted many software companies to develop applications. Microsoft wanted to create an open standard and bring graphical capabilities to any computer that was running MS-DOS.

The first popular graphical operating system came to market in 1984, when Apple released its Macintosh. Everything about the Macintosh's proprietary operating system was graphical, and the Mac was an enormous success. The initial hardware and operating system software Apple released were quite limited but vividly demonstrated the potential of the graphical interface. As the hardware and software improved, that potential was realized.

We worked closely with Apple throughout the development of the Macintosh. Microsoft wanted the Macintosh to sell well and to be widely accepted, not only because we had invested a lot in creating applications for it but also because we wanted the public to accept graphical computing. Steve Jobs led the Macintosh team, and working with him was really fun. Steve has a keen instinct for engineering and design as well as an ability to motivate people that is world class.

It took a lot of imagination to develop graphical computer applications. What should one look like? How should it behave? We inherited some ideas from the work done at Xerox PARC, and some of our ideas were original. At first we went to excess. Our menus, for instance, used nearly every font and icon we could lay our hands on. Then we figured out that all of that busyness made the screen hard to look at, and we developed more sober menus. We created a word processor, Microsoft Word, and a spreadsheet, Microsoft Excel, for the Macintosh. These were Microsoft's first graphical products, and the lessons we learned while creating them proved invaluable to us and many other companies interested in writing applications for the Macintosh and Windows.

The Macintosh had great system software, but Apple refused (until 1995) to let anybody else make computer hardware that would run it.

--------------------------->

*1984: The Apple
Macintosh computer*

This was traditional hardware company thinking: If you wanted to use the software, you had to buy Apple computers.

Whether to keep hardware and software separate was a major issue in the collaboration between IBM and Microsoft to create what was eventually named OS/2, a replacement for MS-DOS, and whether to keep hardware and software standards separate is still an issue today. Some companies treat hardware and software as distinct businesses, and some don't. Software standards create a level playing field for the hardware companies, but many manufacturers use a tie between their hardware and their software to distinguish their systems. These different approaches are already beginning to be played out again in the Internet marketplace.

Throughout the 1980s IBM was an awesome company by every measure capitalism knows. In 1984 it set the record for the most money ever made by any company in a single year—$6.6 billion of profit. In that banner year IBM introduced its second generation personal computer, a high-performance machine called the PC AT, which was built around Intel's 80286 microprocessor ("the 286"). The PC AT was three times faster than the original IBM PC. It was a great success, and within a year it was responsible for more than 70 percent of all business PC sales.

When IBM launched the original PC, it never expected the machine to challenge sales of the company's business systems, although a significant percentage of the PCs were bought by IBM's traditional customers. IBM executives thought the smaller machines would find their market only at the low end. But as microprocessor chips became more powerful, IBM held back on PC development to avoid having PC sales eat into sales of its higher-end products.

In its mainframe business, IBM had always been able to control the adoption of new standards. The company would limit the price/performance ratio (amount of performance available per dollar) of a new line of hardware, for example, so that it wouldn't steal business from existing, more expensive products. It would encourage the adoption of new versions of its operating systems by releasing hardware that required the new operating system software, or it would release operating system software that customers would need new hardware to run. That kind of strategy might have worked well for mainframes, but it was a disaster in the fast-moving personal computer market. IBM could still command somewhat higher prices for performance equivalent to its competitors', but the world had discovered that lots of companies made IBM-compatible hardware and that, if IBM couldn't deliver the right value, someone else would.

Three engineers who appreciated the potential offered by IBM's entry into the personal computer business left their jobs at Texas Instruments and started Compaq Computer. They built personal computers that would accept the same accessory cards as the IBM PC, and they licensed MS-DOS from Microsoft so that their computers could run the same applications as the IBM PC. Compaq produced machines that did everything the IBM PCs did and were more portable too. They embraced the IBM standard and extended it. Compaq quickly became one of the all-time success stories in American business, selling more than $100 million worth of computers its first year. IBM was able to collect royalties by licensing its patent portfolio, but its share of the PC market declined as lower-priced compatible systems came to market and it allowed its PC technology to lag behind.

-------------------------->

To protect the sales of its low-end minicomputers, which weren't much more powerful than a 386-based PC, IBM delayed the release of its PCs based on the powerful Intel 386 chip, Intel's successor to the 286. IBM's delay allowed Compaq to become the first company to bring out a 386-based computer in 1986. This gave Compaq an aura of prestige and leadership that had been IBM's alone.

IBM planned to recover with a one-two punch, the first in hardware and the second in software. It wanted to build computers and write operating systems each of which would depend exclusively on the other for its new features. Competitors would be either frozen out or forced to pay hefty licensing fees. The strategy was to make everybody else's "IBM-compatible" personal computer obsolete.

IBM's plan called for substantial changes in the hardware architecture: new connectors and new standards for accessory cards, keyboards, mice, and even displays. This would raise the compatibility ante. Other PC manufacturers and the makers of peripherals would have to start over, and IBM would have the lead again. To give itself a further advantage, IBM didn't release specifications on any of the new connectors until it had shipped its first systems.

The IBM hardware strategy included some good ideas. One was to simplify the design of the PC by building many features that had formerly been selectable options right into the machine. This would reduce IBM's costs and increase the percentage of IBM components in the ultimate sale. IBM's software strategy was more problematic.

By 1984 a significant part of Microsoft's business was providing MS-DOS to manufacturers who built PCs compatible with IBM's. Around that time we began working independently and then with IBM on OS/2. This time the arrangement wasn't like the MS-DOS agreement. IBM wanted to control the standard in order to help its PC hardware and mainframe businesses, so it became directly involved in the design and implementation of OS/2. Our agreement would allow Microsoft to sell other manufacturers the same operating system that IBM was shipping with its machines, and we would each be allowed to extend the operating system beyond the version we developed together.

OS/2 was central to IBM's corporate software strategy. It was to be the first implementation of IBM's Systems Application Architecture, which the company ultimately intended to be a common development environment across its full line of computers from mainframe to midrange to PC. IBM executives believed that using the company's mainframe technology on the PC would prove irresistible to corporate customers. They thought it would give IBM a huge advantage over PC competitors who didn't have access to mainframe technology. IBM's proprietary extensions to OS/2—called Extended Edition—included communications and database services. IBM also planned to build a full set of office applications—to be called OfficeVision—to work on top of Extended Edition. According to the plan, these applications, including word processing, would make IBM a major player in PC application software, competitive with Lotus and WordPerfect. The development of OfficeVision required another team of thousands. OS/2 was not just an operating system—it was part of a corporate crusade.

Microsoft went ahead and developed OS/2 applications to help get the market going, but as time went on, our confidence eroded. Our development work was burdened by demands that the project meet a variety of conflicting feature requirements as well as by IBM's schedule commitments for Extended Edition and OfficeVision. We had entered into the project in the belief that IBM would allow OS/2 to be enough like Windows that a software developer would have to make at most only minor modifications to get an application running on both operating systems. But after IBM's insistence that applications be compatible with its mainframe and midrange systems, we could see that what we would be left with was more like an unwieldy mainframe operating system than a PC operating system.

Our business relationship with IBM was vital to us. In 1986 we had taken Microsoft public to provide liquidity for employees who had been given stock options. It was about that time that Steve Ballmer and I proposed to IBM that it buy up to 30 percent of Microsoft—at a bargain price—so that it would share in our fortune, good or bad. We thought this might help the companies work together more amicably and productively. IBM wasn't interested.

We worked extremely hard to make our operating system collaboration with IBM succeed. I felt that the OS/2 project would be a ticket to the future for both companies. Instead, it eventually created an enormous rift between us.

A new operating system is a big project, and this one was far-flung. We had our team working outside Seattle. IBM had teams in Boca Raton, Florida; Hursley Park, England; and later, Austin, Texas.

But the geographical distribution of the team didn't pose as big a problem for OS/2 development as the problems that came from IBM's mainframe legacy. IBM's software projects had almost never caught on with PC customers precisely because they had been designed with a mainframe customer in mind. For instance, it took three minutes for one version of OS/2 to "boot" (to make itself ready for use after it was turned on). Three minutes didn't seem bad to IBM because, in the mainframe world, booting could take fifteen minutes.

With more than 300,000 employees, IBM was also stymied by its commitment to company-wide consensus. Every part of IBM was invited to submit Design Change Requests, which usually turned outto be demands that the personal computer operating system design be changed to better fit the needs of mainframe products. Our combined teams got more than 10,000 such requests, and talented people from IBM and Microsoft would sit and discuss them for days.

I remember change request #221: "Remove fonts from product. Reason: Enhancement to product's substance." Someone at IBM didn't want the PC operating system to offer multiple typefaces because a particular IBM mainframe printer couldn't handle them.

Finally it became clear to us that joint development wasn't going to work. We asked IBM to let us develop the new operating system on our own and license it to them cheaply. We'd make our profit by selling the same operating system to other computer companies. But IBM had declared that its own programmers had to be involved in the creation of any software it considered strategic. And operating system software was clearly that.

IBM was such a great company. Why did it have so much trouble with PC software development? One answer is that IBM tended to promote all of its good programmers into management and leave the less talented behind. Even more significant, IBM was haunted by its successful past. Its traditional, hardware-centric engineering process was unsuitable for the rapid pace and market requirements of PC software.

In April 1987 IBM released its one-two punch, which was supposed to beat back the imitators. The "clone-killer" hardware was called the PS/2, and it ran the new operating system, OS/2.

The PS/2 included a number of innovations. The most celebrated was "Microchannel bus" circuitry, intended to be an improved way for accessory cards to connect to the system. Accessory cards permitted PC hardware to be extended to meet particular customer requirements such as sound or mainframe communications capabilities. The PS/2's Microchannel bus was an elegant replacement for the connection bus in the PC AT. It was potentially much faster than the PC AT's bus. But in actual practice the speed of the old bus hadn't been holding anyone up, so the Microchannel bus solved problems that most customers didn't have. More important, the Microchannel didn't work with any of the thousands of add-in cards that worked with the PC AT and compatible PCs.

Eventually IBM agreed to license the Microchannel architecture for a royalty to manufacturers of add-in cards and PCs. But by then a coalition of manufacturers had already announced a new bus with many of the capabilities of the Microchannel bus but compatible with the PC AT bus. Customers rejected the Microchannel machines in favor of machines equipped with the PC AT-compatible bus, which forced IBM to continue to release machines that supported the AT-compatible bus. The real casualty of the bus war was IBM's loss of control over the personal computer's architecture. Never again would it be able to single-handedly move the industry to a new design.

Despite a great deal of promotion from both IBM and Microsoft, customers thought OS/2 was too unwieldy and complicated. The worse OS/2 looked to us, the better Windows looked. We'd lost the chance to make Windows and OS/2 compatible, and because we'd lost the struggle to make OS/2 run on modest machines, it only made sense to continue to develop Windows. Windows was far "smaller"—meaning it used less hard disk space and could work in a machine with less memory, so there would be a place for it on machines that could never run OS/2. We called this our "family" strategy: OS/2 would be the high-end system, and Windows would be the junior member of the family, for smaller machines.

IBM was never happy about our family strategy, but it had its own plans. In the spring of 1988 it joined other computer makers in establishing the Open Software Foundation(OSF) to promote UNIX, an operating system that had originally been developed at AT&T's Bell Labs in 1969 but that over the years had splintered into a number of versions. Some of the versions of UNIX were developed at universities, where UNIX was used as a working laboratory for operating systems theory. Other versions were developed by computer companies. Each company enhanced UNIX for its own computers, which made its operating system incompatible with everyone else's. UNIX had evolved from a single system into a collection of operating systems competing with each other. The many differences between versions had made software compatibility harder and had held back the rise of a strong third party software development effort for UNIX. Only a few software companies could afford to develop and test applications for a dozen different versions of UNIX. And computer software stores couldn't afford to stock all the different versions.

The Open Software Foundation was the most promising of several attempts to "unify" UNIX by creating a common software architecture that would work on various manufacturers' hardware. In theory, a unified UNIX could get a positive-feedback cycle going, and that's what its backers hoped for. But despite significant funding, it turned out to be impossible for the Open Software Foundation to mandate cooperation from a committee of vendors who were competing for every sale. OSF members, including IBM, DEC, and others, continued to promote the benefits of their particular versions of UNIX. Each UNIX company tried to suggest that its system was superior. But if you bought a UNIX system from one vendor, your software couldn't automatically run on any other UNIX system. You were tied to that vendor, whereas in the PC world you had a choice of companies from which to buy your hardware.

The problems of the Open Software Foundation initiative and similar initiatives point up the difficulty of trying to impose standards in a field in which innovation is moving rapidly and the companies that make up the standards committee are competitors. The marketplace in computers or consumer electronics adopts standards because customers insist on them. Standards ensure interoperability, minimize user training, and foster the largest possible range of software choices. Any company that wants to create a standard has to price its products very reasonably, or the standard won't be adopted. The market effectively chooses a standard whose products are reasonably priced and replaces it when its products are obsolete or too expensive.

Microsoft operating systems are offered today by over 1,000 different manufacturers, which gives customers an immense range of choices and options. Microsoft has been able to provide compatibility because hardware manufacturers have agreed not to allow modifications to our software that would introduce incompatibility. The hundreds of thousands of software developers don't need to worry about what PCs their software will run on. Although the term "open" is used in many different ways, to me it means offering hardware choices and software applications choices to the customer.

Consumer electronics products have also benefited from standards managed by private companies. Years ago consumer electronics companies often tried to restrict competitors from using their technologies, but now all of the major consumer electronics makers are quite open to licensing their patents and trade secrets. The royalties are typically under 5 percent of the cost of the device. Audiocassettes, VHS tapes, compact discs, televisions, and cellular telephones are all examples of technologies that were created by private companies that receive royalties from everyone who makes the equipment. Dolby Laboratories' algorithms, for example, are the de facto standard for noise reduction.

In May 1990, the last weeks before the release of Windows 3.0, we tried to reach an agreement with IBM for it to license Windows to use on its personal computers. We told IBM we thought that, although OS/2 would work out over time, for the moment Windows was going to be a success and OS/2 would find its niche slowly. IBM initially declined to license Windows, although later it changed its mind.

In 1992 IBM and Microsoft stopped their joint development of OS/2. IBM continued to develop the operating system alone. The ambitious plan for the OfficeVision applications was eventually canceled.

Analysts estimate that IBM poured more than $2 billion into OS/2, OfficeVision, and related projects. If IBM and Microsoft had found a way to work together, thousands of people-years—the best years of some of the best employees at both companies—would not have been wasted. If OS/2 and Windows had been compatible, graphical computing would have become mainstream years before it finally did.

The acceptance of graphical interfaces was also held back because most major software applications companies didn't invest in them. They largely ignored the Macintosh and ignored or ridiculed Windows. Lotus and WordPerfect, the market leaders for spreadsheet and word-processing applications, made only modest efforts on OS/2. In retrospect, ignoring the graphical interface for so long was a costly mistake for them. When Windows finally benefited from a positive-feedback cycle, generated by applications from many of the small software companies as well as from Microsoft, the former market leaders fell behind.

The success of Windows was a long time coming. For most of the 1980s many people had written it off because it required any computer running it to have lots of expensive memory and it demanded more of applications. "Why do we want another layer of software on top of MS-DOS?" critics asked. "Won't that slow things down?" "Why would people be willing to sacrifice performance for graphics?" It took many years of persistence, a very long-term approach, to make Windows a success.

Windows, like the PC, continues to evolve. Microsoft continues to enhance it, and anyone can develop applications software that runs on the Windows operating system without having to notify or get permission from Microsoft. In fact, today there are tens of thousands of commercially available software packages for Windows, including offerings that compete with most Microsoft applications.

Customers tell me they worry that Microsoft, by definition the only source for Microsoft operating system software, could raise prices or slow down or even stop innovation. But if we did, we wouldn't be able to sell our new versions. Existing users wouldn't upgrade, and we wouldn't get any new users. Our revenue would fall, and other companies would come in and take our place. The positive-feedback mechanism helps challengers as well as the incumbent. A leader can't rest on its laurels because there's always a competitor coming up from behind.

Recently it's been fashionable to say both that Microsoft is unstoppable and that its days of leadership are numbered. Both views can't be correct. Certainly, Microsoft is not unstoppable. We have to earn our leadership position every day. If we stop innovating or stop adjusting our plans, or if we miss the next big turn in the industry's road, we'll lose out. Because Microsoft isn't immune to failure, we're careful not to dismiss the doomsayers. We heed their warnings, and we ask: "Why are they saying that? Are we being critical enough of ourselves? Are we missing a new technology?"

No product stays on top unless its company keeps improving it, and we've had to keep innovating. The Macintosh software might have become the successor to MS-DOS. OS/2 or UNIX might have. As it happened, MS-DOS was succeeded by Windows, and each subsequent version of Windows will be successful with new users only if current users adopt it. Microsoft has to do its best to make new versions so attractive in terms of price and features that people will want to change even though it involves overhead for both developers and customers. Only a major advance can convince enough users it's worth their while to make a change. With sufficient innovation, the case to consumers can be made. I expect major new generations of Windows to come along every two to three years.

The continued success of Windows is by no means guaranteed. We've had to improve our operating system software to keep up with advances in both computer hardware and the communications infrastructure, and the seeds of new competition are being sown constantly in research environments and garages around the world. Sony is entering the market with an operating system aimed at products that combine computers with consumer electronics gear. Netscape plans to make its Internet browser software evolve into a kind of operating system. Sun Microsystems wants to make the programming routines in its Java language runtime library numerous enough and capable enough to act as an operating system. Today's main challenge for us is making Windows the best way to gain access to the Internet, although there are other important initiatives too. Three-dimensional graphics are becoming important, and when speech- and handwriting-recognition become accurate enough, these innovations will cause another big change in operating systems.

-------------------------➤

People ask, "Will the Internet be the thing that kills you?"

I say, with tongue in cheek, "No. It's all the other things that will kill us, because we're so focused on the Internet."

The Internet is changing the rules for everybody in the software industry. Novell made a name in network software, but internal, private corporate Internets—called "intranets"—are the wave of the future. As the popularity of 1-2-3 waned, Lotus pinned its hopes on a networking application called Notes. IBM bought Lotus in 1995 so that it could own Notes, but now the software faces competition from Microsoft Exchange and other Internet-based applications. Under Lou Gerstner's leadership, IBM has become far more efficient and has regained both its profitability and its positive focus on the future. IBM's server business shows renewed health. Although the long-term decline in mainframe revenues remains a problem for IBM, it will clearly be one of the major companies providing products for businesses and the Internet.

Our business moves too fast to spend much time looking back. I pay close attention to our mistakes, however, as I try to focus on future opportunities. It's important to acknowledge mistakes and to make sure you draw the right lessons from them. It's also important to make sure that nobody avoids trying something new because he thinks he'll be penalized for what happens if it doesn't work out. Almost no single mistake is fatal.

In recent years Microsoft has deliberatcly hired a few managers with experience in failing companies. When you're failing, you're forced to be creative, to dig deep and think, night and day. I want some people around who have been through that. Microsoft is bound to have failures in the future, and I want people here who have proved they can do well in tough situations.

Death can come swiftly to a market leader. By the time you're thrown out of the positive-feedback cycle, it can be too late to change what you've been doing, and all the elements of a negative spiral can come into play. It's difficult to recognize that you're in a crisis and to react to it when your business seems to be extremely healthy.

The need to look down the road keeps me alert. I never anticipated Microsoft's growing so big, and now, at the beginning of this new era, I unexpectedly find myself a part of the establishment. My goal is to prove that a successful corporation can renew itself and stay in the forefront.

4

INFORMATION APPLIANCES AND APPLICATIONS

When I was a kid, *The Ed Sullivan Show* came on at eight o'clock on Sunday nights. Most Americans with television sets tried to be at home to watch it because that might be the only time and place to see the Beatles, Elvis Presley, the Temptations, or that guy who could spin ten plates simultaneously on the noses of ten dogs. But if you were driving back from your grandparents' house or on a Cub Scout camping trip, too bad. Not being at home on Sunday at eight meant that you also missed out on the Monday morning talk about Sunday night's show.

Conventional television allows us to decide what we watch but not when we watch it. The technical term for this sort of broadcasting is "synchronous." Viewers have to synchronize their schedules with the time of a broadcast that's sent to everybody at the same time. That's how I watched *The Ed Sullivan Show* thirty years ago, and it's how most of us will watch the news tonight.

In the early 1980s the videocassette recorder gave us more flexibility. If you cared enough about a program to fuss with timers and tapes in advance, you could watch it whenever you liked. You could claim from the broadcasters the freedom and luxury to serve as your own program scheduler—and millions of people do. When you tape a television show, or when you let your answering machine take an incoming message so that you don't have to pick up the phone, you're converting synchronous communications into a more convenient form: "asynchronous" communications.

It's human nature to find ways to convert synchronous communications into asynchronous forms. Before the invention of writing 5,000 years ago, the only form of communication was the spoken word and the listener had to be in the presence of the speaker or miss his message. Once the message could be written, it could be stored and read later by anybody, at his or her convenience. I'm writing these words at home on a summer evening, but I have no idea where or when you'll read them. One of the benefits the communications revolution will bring to all of us is more control over our schedules.

Once a form of communication is asynchronous, you also get an increase in the variety of selection possibilities. Even people who rarely record television programs routinely rent movies from the thousands of choices available at local video rental stores for just a few dollars each. The home viewer can spend any evening with Elvis, the Beatles—or Greta Garbo.

Television has been around for fewer than sixty years, but in that time it has become a major influence in the life of almost everyone in the developed nations. In some ways, though, television was just an enhancement of commercial radio, which had been bringing electronic entertainment into homes for twenty years. But no broadcast medium we have right now is comparable to the communications media we'll have once the Internet evolves to the point at which it has the broadband capacity necessary to carry high-quality video.

Because consumers already understand the value of movies and are used to paying to watch them, video-on-demand is an obvious development. There won't be any intermediary VCR. You'll simply select what you want from countless available programs.

No one knows when residential broadband networks capable of supporting video-on-demand will be available in the United States and other developed countries, let alone in developing countries. Many corporate networks already have enough bandwidth, but as I'll explain in chapter 5, even in the U.S. most homes will have to make do for some time—maybe more than a decade—with narrowband and midband access. Fortunately, these lower-capacity bandwidths work fine for many Internet-based services such as games, electronic mail, and banking. For the next few years, interactivity in homes will be limited to these kinds of services, which will be delivered to personal computers and other information appliances.

Even after broadband residential networks have become common, television shows will continue to be broadcast as they are today, for synchronous consumption. But after they air, these shows—as well as thousands of movies and virtually all other kinds of video—will also be available whenever you want to view them. If a new episode of *Seinfeld* is on at 9:00 P.M. on Thursday night, you'll also be able to see it at 9:13 P.M., 9:45 P.M., or 11:00 A.M. on Saturday. And there will be thousands of other choices. Your request for a specific movie or TV show episode will register, and the bits will be routed to you across the network. It will feel as if there's no intermediary machinery between you and the object of your interest. You'll indicate what you want, and presto! you'll get it.

Movies, TV shows, and other kinds of digital information will be stored on "servers," which are computers with capacious disks. Servers

1995: A personal computer–based interactive media server

will provide information for use anywhere on the network, just as they do for today's Internet. If you ask to see a particular movie, check a fact, or retrieve your electronic mail, your request will be routed by switches to the server or servers storing that information. You won't know whether the movie, TV show, query response, or e-mail that arrives at your house is stored on a server down the road or on the other side of the country, and it won't matter to you.

The digitized data will be retrieved from the server and routed by switches back to your television, personal computer, or telephone—your "information appliance." These digital devices will succeed for the same reason their analog precursors did—they'll make some aspect of life easier. Unlike the dedicated word processors that brought the first microprocessors to many offices, most of these information appliances will be general-purpose, programmable computers connected to the network.

Even if a show is being broadcast live, you'll be able to use your infrared remote control to start it, stop it, or go to any earlier part of the program, at any time. If somebody comes to the door, you'll be able to pause the program for as long as you like. You'll be in absolute control—except, of course, you won't be able to forward past part of a live show as it's taking place.

Most viewers can appreciate the benefits of video-on-demand and will welcome the convenience it gives them. Once the costs to build a broadband network are low enough, video-on-demand has the potential to be what in computer parlance is called a "killer application," or just "killer app"—a use of technology so attractive to consumers that it fuels market forces and makes the underlying invention on which it depends all but indispensable. Killer applications change technological advances from curiosities into moneymaking essentials.

The term "killer application" is relatively new, but the idea isn't. Thomas Edison was as great a business leader as he was an inventor. When he founded the Edison General Electric Company in 1878, he understood that to sell electricity he had to demonstrate its value to consumers. Edison lit up the public's imagination with the promise that electric lighting would become so cheap that only the rich would buy candles. He correctly foresaw that people would pay to bring electric power into their homes so that they could enjoy a great application of electric technology—light.

A number of additional applications for electricity quickly found popular acceptance. The Hoover Company sold an electric sweeping machine, and soon there were electric stoves, heaters, toasters, refrigerators, washing machines, irons, power tools, hair dryers, and a host of other laborsaving appliances. Electricity became a basic utility.

Sometimes an application that ends up being a killer application wasn't anticipated by the product's inventor. Avon Skin-So-Soft was just another lotion competing in a crowded market until somebody discovered that it repelled insects. Now it may still be sold for its original application—to soften skin—but its sales have increased because of its new application. And when Tim Berners-Lee invented the World Wide Web in 1989 as a way for high energy physicists to exchange information, he didn't foresee all the great applications that would be developed for it.

In the late 1970s the market for dedicated word processors grew incredibly fast, until it included more than fifty manufacturers with combined sales of more than $1 billion annually. Word processing was a killer application of microprocessor technology. But within a couple of years personal computers appeared, and their ability to run different types of applications was something new. A PC user could quit Word-Star (for years one of the most popular word processing applications) and start up another application, such as the spreadsheet program Visi-Calc or the database management program dBASE. Collectively, Word-Star, VisiCalc, and dBASE were attractive enough to motivate the purchase of a personal computer. They were killer applications for the first popular PCs.

The first killer application for the original IBM PC was Lotus 1-2-3, a spreadsheet tailored to the strengths of that machine. The Apple Macintosh's killer business applications were Aldus PageMaker for designing documents to be printed, Microsoft Word for word processing, and Microsoft Excel for spreadsheets. Early on, more than a third of the Macintoshes used in business and many of the Macs in homes were bought for one killer application—what became known as desktop publishing.

I use the term "application" quite broadly here. Microsoft Excel is an application, but so is home shopping or videoconferencing. Any type of

World Wide Web use is an application, but no specific use is a killer application yet because none of them alone is driving large numbers of people to hook up to the Internet. The attraction of the Web is its breadth of uses and, in its early days at least, its novelty. But information retrieval, education, entertainment, shopping, and e-mail will in time become killer applications for the Internet.

Allowing people to connect to the Internet through a variety of information appliances will be critical to making Internet use a mainstream activity. In the years ahead we'll see a proliferation of digital devices that will take on different forms and communicate at different speeds, enabling each of us to stay in touch over the net with other people as well as with information of all kinds. We'll use new versions of familiar tools—telephones, TVs, PCs, white boards, notebooks, wallets—to take command of information and reshape the media that make up much of our daily life: books, magazines, newspapers, video, music, telephones, games, even the art on the walls. We don't know exactly what all of the successful appliances will look like, but we do know that as technology evolves an increasing number of the appliances will be general-purpose, programmable computers connected visibly or invisibly to the net.

This idea that general-purpose computers will prevail deserves more attention. When the multifaceted PC won the word processing market away from Wang's dedicated machines, it was the triumph of the general-purpose microcomputer over the special-purpose microcomputer. A "word processor" was no longer a physical machine; it was a software application. The telephone answering machine underwent the same transformation, evolving into voicemail, a software application running on a computer (often at a telephone company's central office). When video-on-demand finally becomes available, the home movie player will undergo the same evolution, from a special-purpose tape recorder (VCR) to a software application running on a computer connected to a communications network.

There are many other examples of the evolution of a tool from special-purpose hardware into software running on general-purpose hardware. The first wave of the microprocessor revolution brought us machines that imitated older tools: electronic cash registers replaced mechanical cash registers, electronic calculators replaced adding machines, and electronic games replaced a lot of physical objects including playing cards and pinball machines. In the second wave, microprocessors came into their own as the hearts of general-purpose PCs that made special-purpose machines unnecessary. Thanks to software, a single machine could assume many guises, with consumers benefiting from the consequent competition and mass-market economies. Software applications taught computers to function as cash registers, calculators, and games, not to mention word processors, spreadsheets, databases, and telephone answering systems. A general-purpose computing machine, the dream of Charles Babbage, could do it all.

As the Internet matures, there will be dozens of new consumer machines, many of them devoted to a single purpose. Home audio gear will evolve. New kinds of game machines will come and go, with many of the most popular models featuring modems and software that will let players connect up to the Internet and play or browse from different locations. High-end telephones will have screens that display information, including yellow pages advertising. Various companies will promote terminals specifically for Web browsing. Cellular phones and pagers will get more powerful. Some of these special-purpose devices will find a place in the market for a few years, but in the long run almost all of them will give way to programmable, general-purpose devices— "computers"—connected visibly or invisibly to the network.

This evolution from special-purpose devices to general-purpose devices will be apparent in the successive generations of set-top boxes that will connect TV sets to networks. Conventional set-top boxes are tuners that receive analog signals from as many as dozens of cable channels and pass signals from one of the channels at a time on to your television set. If you've paid to watch channels that are scrambled, the set-top box unscrambles them. A generation of set top boxes reaching the mass market in about 1997 will have the additional ability to handle compressed digital video signals, which will substantially increase the number of channels that can be received and also support Internet browsing. Eventually your television set will connect to the net via a new set-top box that will be a very powerful general-purpose computer—and might not be set on the top of the TV at all. It might be located inside a television, behind a television, on top of a television, on a basement wall, or even outside the house. Your TV, like your PC, will connect to the network and conduct a "dialogue" with the net's switches and servers, relaying your choices and retrieving information and programming.

Personal computers will continue to evolve, getting easier to use and less expensive. Within a few years, many people won't think of a home PC as a computer as much as they'll think of it as a simple tool for accomplishing a number of tasks, including entertainment tasks.

However much like a PC the TV becomes, and vice versa, there will continue to be a critical difference between the way a PC is used and the way a TV is used: the distance from which they are viewed. Today more than a third of U.S. households have personal computers (not counting game machines). Eventually almost every home will have at least one PC connected directly to the net. This is the appliance you'll use when details count or when you want to type. It will put a high-quality monitor a foot or two from your face so that your eyes can focus easily on text and other small images. A big-screen TV across the room doesn't lend itself to the use of a keyboard, and it doesn't give you any privacy, although it's ideal for applications that multiple people watch at the same time.

Set-top boxes and PC-interface equipment will be designed so that

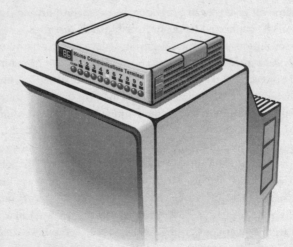

Prototype sketch of a television set-top box

even the oldest television sets as well as the most current personal computers can be used on the interactive network, but new TVs and PCs will offer better pictures. The quality of the images on today's TV sets is quite poor compared to the quality of pictures in magazines or the quality of the images on movie theater screens. While U.S. television signals can transmit 486 lines of picture information, not all of those lines are distinguishable on most sets, and the typical home VCR can record or play back only about 280 lines of resolution. That's why it's so difficult to read the credits at the end of a movie on a television set. Conventional television screens are also a different shape from most movie theater screens. Our TVs have an "aspect ratio" (the relationship of picture width to height) of 4 to 3, meaning that a picture is somewhat wider than it is tall. Feature films are typically made with an aspect ratio of about 2 to 1—twice as wide as they are tall.

High-definition television (HDTV) that offers more than 1,000 lines of resolution, with better color and a 16 to 9 aspect ratio, has been demonstrated, and it's beautiful to see. But despite the efforts of both government and industry in Japan, where the analog technology was created, HDTV didn't catch on because it required expensive new equipment for both broadcasting and receiving. Advertisers wouldn't pay extra to fund HDTV because it doesn't make ads measurably more effective. However, digital forms of HDTV might still catch on because the broadband network will allow video to be received at multiple resolutions and aspect ratios. The idea of adjustable resolution is familiar to users of personal computers, who can choose 480 lines of resolution (called VGA) or higher resolutions of 600, 768, 1,024, or 1,200 horizontal lines, depending on what their monitors and display cards can support.

Both TV screens and PC screens will continue to improve in quality. Most will be flat-panel displays. A new kind of screen will be the digital white board: a large wall-mounted screen, perhaps an inch thick, that will take the place of today's blackboards and white boards. The digital white board will display pictures, movies, and other graphical information, as well as text and other fine details. People will be able to draw or write notes and lists on it. The computer controlling the white board will recognize handwritten notes or lists and convert the handwriting into a readable typeface. The digital white board will show up first in conference rooms and then in private offices and even homes.

Telephones will connect to the same networks as the PCs and TVs. Many phones will have small, flat screens and tiny cameras. Otherwise, though, they'll look more or less like today's phones. Kitchens will continue to have wall phones that conserve counter space. You'll sit close to the phone and look at a screen that shows the person you're talking to— or at a stock picture he or she transmits instead of live video. Technologically, the phone hanging over a dishwasher tomorrow will have a lot in common with the set-top box in the living room and the personal computer in the den, even though it will assume the form of a phone. Under the hood, many of the information appliances will have pretty much the same computer architecture.

In a mobile society, people need to be able to work efficiently while

1996: Multimedia laptop computer from Digital Equipment Corporation shown with a personal computer and a palmtop device

they're on the road. Two centuries ago a traveler could carry a "lap desk," a hinged writing board attached to a thin mahogany box with a drawer for pens and ink. When folded, it was reasonably compact, and when opened, it provided an ample writing surface. *The Declaration of Independence* was written on a lap desk in Philadelphia, a long way from Thomas Jefferson's Virginia home. The need for a portable writing station is met today by the laptop, a folding, lap-size personal computer. Many people who work from both office and home—including me—choose a laptop (or a slightly smaller computer, known as a notebook) as their primary computer. In the office, the small computer can be connected to a large monitor and to the corporate network. Notebook computers will continue to get thinner until they're about the size of a tablet of paper.

Notebooks are the smallest and most portable real computers today, but we'll soon have pocket-size computers with snapshot-size color screens. When you whip one out, nobody will say, "Wow! You've got a computer!"

What do you carry on your person now? Probably at least keys, identification, money, and a watch. And maybe credit cards, a checkbook, traveler's checks, an address book, an appointment book, a notepad, something to read, a camera, a pocket tape recorder, a cellular phone, a pager, concert tickets, a map, a compass, a calculator, an electronic entry card, photographs, and maybe a loud whistle to call for help.

You'll be able to keep equivalent necessities—and more—in an information appliance I call the wallet PC. It will be about the same size as a wallet, which means you'll be able to carry it in your pocket or purse. It will display messages and schedules and let you read or send electronic mail and faxes, monitor weather and stock reports, and play both simple and sophisticated games. At a meeting, you might take notes, check your appointments, browse information if you're bored, or choose from among thousands of easy-to-call-up photos of your kids.

Rather than hold paper currency, the new wallet will store unforgeable digital money. Today when you hand somebody a dollar bill, a check, a gift certificate, or some other negotiable instrument, the transfer of paper represents a transfer of funds. But money doesn't have to be expressed on paper. Credit card charges and wired funds are exchanges of digital financial information. Tomorrow the wallet PC will make it easy for anyone to spend and accept digital funds. Your wallet will link into a store's computer to allow money to be transferred without any physical exchange at a cash register. Digital cash will be used in interpersonal transactions too. If your son needs money, you might slip five bucks from your wallet PC into his digitally.

When wallet PCs have become ubiquitous, we can eliminate the bottlenecks that plague airport terminals, theaters, and other places where people queue up to show their identification or a ticket. As you pass through an airport gate, for example, your wallet PC will connect to the airport's computers and verify that you've paid for a ticket. You won't need a key or a magnetic card key to get through doors either. Your wallet PC will identify you to the computer controlling the lock.

As cash and credit cards begin to disappear, criminals may target wallet PCs, so there will have to be safeguards to prevent a wallet PC from being used in the same way a stolen credit card can be. The wallet PC will store the "keys" you'll use to identify yourself. You'll be able to invalidate your keys easily, and you'll be able to change them regularly. You might have to enter a password at the time of important transactions. Automatic teller machines ask you to provide a personal identification number, just a very short password, now. An option that would eliminate the need for people to remember passwords would be using biometric measurements—such as voiceprints or fingerprints—for security. A person's biometric measurements are more secure than passwords and will almost certainly protect some wallet PCs.

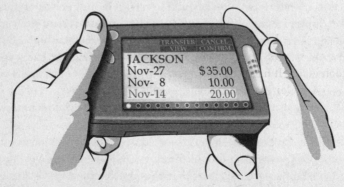

Prototype sketch of a wallet PC

In a biometric security system, your wallet PC might demand that you read out loud a random word it flashes on its screen or that you press your thumb against its side whenever you're about to conduct a transaction that has significant financial implications. The wallet will compare what it "heard" or "felt" with its digital record of your voice-print or thumbprint.

Wallet PCs with the right equipment will be able to tell you exactly where you are anyplace on the face of the earth. The Global Positioning System (GPS) satellites that orbit Earth right now broadcast signals that enable jetliners, oceangoing boats, cruise missiles, some cars—and even hikers with handheld GPS receivers—to know their exact locations. Such devices are currently available for a few hundred dollars, and eventually they'll be built into many wallet PCs.

The wallet PC will connect you to the interactive network while you travel and tell you where you are. A voice from its built-in speaker will let you know that a freeway exit is coming up or that the next intersection has frequent accidents. It will monitor digital traffic reports and warn you that you'd better leave for the airport early, or it will suggest an alternative route. The wallet PC's color maps will overlay your location with whatever kinds of information you want—road and weather conditions, campgrounds, scenic spots, even fast-food places. You might ask, "Where's the closest Chinese restaurant that's still open?" and the answer will be transmitted to your wallet by wireless network. Off the roads, on a hike in the woods, the wallet PC will be your compass and as useful as your Swiss Army knife.

In fact, I think of the wallet PC as the new Swiss Army knife. I had one of those knives when I was a kid. Mine wasn't the most basic, with just two blades, but it wasn't one of those with a workshop's worth of tools either. It had the classic shiny red handle with the white cross and lots of blades and attachments, including a screwdriver, a tiny pair of scissors, and even a corkscrew (although at the time I had no use for that particular accessory). Some wallet PCs will be simple and elegant and offer only the essentials—say, a small screen, a microphone, a secure way to transact business with digital money, and the capability to read or otherwise use basic information. Other wallet PCs will bristle with all kinds of gadgets—cameras, scanners that will be able to read type or handwriting, and receivers with the global positioning capability. Most wallet PCs will have a panic button you can press if you need emergency help. Some models will include thermometers, barometers, altimeters, and heart-rate sensors.

Precursors of the wallet PC are on the market already. In addition to cellphones and pagers, hand-held computers pack a number of functions. Personal digital assistants (PDAs) are available from Sharp, Hewlett-Packard, PSION, and others. Some of the first generation PDAs were underpowered and didn't have software sophisticated enough to interest a broad set of users. Apple's Newton relied on handwriting and didn't live up to the high expectations Apple set for it. Microsoft didn't even bring its software for first generation PDAs to market. A better generation of PDAs is reaching consumers beginning in 1997, and I expect a big market to develop over time.

The simplest precursor of the wallet PC, already popular in Europe, is the so-called smart card. It resembles a credit card but has a microprocessor embedded right inside its plastic. The smart card of the future will identify its owner and store digital money, tickets, and medical information. It won't have a screen, audio capabilities, or any of the elaborate options of the more expensive wallet PCs. It will be handy for travel or as a backup to a wallet PC, and it may be sufficient by itself for some people's uses.

Prices for wallet PCs will vary widely, just as the prices of cameras vary today. Simple, single-purpose smart cards for digital currency will cost about what a disposable camera does now, whereas a really sophisticated wallet PC might cost what an elaborate camera does now, $1,000 or more.

If you aren't carrying a wallet PC, you'll still have access to the interactive network at kiosks—some free, some requiring payment of a fee—in office buildings, shopping malls, and airports, in much the same way that drinking fountains, rest rooms, and pay phones are available now. These kiosks will replace not only pay phones but also banking machines. They'll offer many other network applications, from sending and receiving messages to buying tickets. Some kiosks will display advertising links to specific services when you first log on—the way phones in airports connect you directly to hotel and rental car desks now. Kiosks will be rugged devices on the outside and PCs on the inside.

No matter what form the PC takes, from wallet to desktop to kiosk, users will still have to be able to navigate their way through its applications. Today's PCs and the Internet don't make this easy enough. Neither do television remote controls. Future systems with more choices will have to do better than make you go step-by-step through all the options. Instead of having to remember the channel of a TV program, you'll have the ability to use a graphical menu that will let you select what you want by pointing to an easy-to-understand image.

Eventually we'll also be able to speak to televisions, personal computers, or other information appliances. At first we'll have to stick to a limited vocabulary, but eventually our exchanges with our appliances will become quite conversational. The speech recognition capability requires powerful hardware and software because talk that a human can understand effortlessly is very hard for a computer to interpret. Speech recognition works fine already for a small set of predefined commands, such as "Call my sister," but it's much more difficult for a computer to decipher a sentence it isn't prepared for. Within the next ten years computers will do a far better job, in part because they'll read your lips as well as listen to you speak (it's easier to understand someone if you can watch his lips move).

Lip reading will be one benefit of the video cameras that will become standard PC equipment once videoconferencing is popular. A camera will also let the PC recognize who is using it so that the PC can better anticipate the person's needs or carry out policies. For example, a PC might refuse to respond to a person it didn't know, or it might decline to connect to an adult Web site if it "saw" that a child was nearby. Video cameras will also allow "gesture input." When you nod or shake your head, turn thumbs down, or wave good-bye, your PC will know what you mean.

Some users will find it convenient to handwrite instructions to a computer. Many companies, including Microsoft, have spent some years working on what we call "pen-based computers" that are capable of reading handwriting. I was overly optimistic about how quickly we'd be able to create software that would recognize the handwriting of a broad range of people. The difficulties turned out to be quite subtle. When we tested the system ourselves, it worked well, but new users continued to

-------------------------------▶

have trouble with it. We discovered that we were unconsciously making our handwriting neater and more recognizable than usual. We were adapting to the machine rather than demanding that the machine adapt to us. Once, when the development team thought they had finally created a program that worked, they rushed over to my office to demonstrate their achievement. But what had been working for them didn't work for me. It turned out that everybody on the project happened to be right-handed, and the computer, which was programmed to look at the strokes in the writing, couldn't interpret the very different strokes in my left-handed penmanship. Getting a computer to recognize handwriting is as difficult as getting one to recognize speech. But even though the challenges have proved greater than I believed they were at first, I remain confident that we can find a solution and that the ultimate market for pen-based computers will be huge. In the future lots of people will be taking handwritten notes on computer tablets rather than paper.

Whether you give a command by voice, in writing, or by pointing, you probably won't stand for being confused or frustrated or for having your time wasted. The interactive network's software will have to make it almost infallibly easy to find information, to navigate, even when users don't know exactly what they're looking for.

"Information overload" will be a real concern to people who imagine that the fiber-optic cables of the broadband interactive network will be like enormous pipes spewing out large quantities of information. Information overload isn't unique to the Internet, though, and it doesn't have to be a problem. We already cope with astonishing amounts of information by relying on an extensive infrastructure that has evolved to help us be selective—everything from library catalogs, movie reviews, and the yellow pages to recommendations from friends. When people say they're worried about information overload, ask them to consider how they choose what to read. When we visit a bookstore or a library, we don't worry about reading every volume. Navigational aids point us to information we're interested in and help us find the books and magazines we want. We read book reviews, and we use the library's Dewey decimal system. Once we have a book in hand, we may use its table of contents and index to navigate to the specific information we want.

Approximations of the table of contents and the index have become wildly popular navigation aids on the Internet's World Wide Web. A service such as YAHOO (http://www.yahoo.com) plays the role of a vast table of contents to the thousands of Web sites worldwide. Researchers at YAHOO read and catalog Web pages constantly. When you visit the site, you select a general category and then progressively refine your selections. If you want information about the television show *Friends*, you choose the broad category of Entertainment and then work your way down through subcategories: Television, Shows, Comedies, and then Friends. On the final page you see a list of Web pages devoted to the show. Click on the name of one of the pages, and you connect to it. With all the information on the Internet, it's nice to have a human sensibility there to help you locate Web pages of interest.

A service such as AltaVista (http://altavista.digital.com) offers an index of the Web's contents. Computers at the service, which is run by Digital Equipment Corporation, systematically and continuously browse millions of Web pages, recording in a vast index information about the interesting words found on each page. When you visit the site, you type in one or more words that interest you, and AltaVista searches its index and compiles a list of all sites that have words matching your criteria. I have fun with AltaVista because it lets me find a needle in a haystack, but it's also frustrating to use because it has no editorial judgment. I just used AltaVista to search for "Friends," and a list of 100,000 sites came back in no particular order. The first was a map showing how to get to the Cambridge Friends School near Boston, the second was the home page of a jewelry store called Gems for Friends, the third was a list of the friends of a 19-year-old student in Uppsala, Sweden, and so on. This isn't much help. But if I wanted to find information about a particular friend, AltaVista would be great because it indexes names.

Both of these types of search services have generated a lot of interest in the financial communities because they meet an obvious need—the need to find things on a network where the growth in the amount of information is staggering. I suspect that a large number of additional services will appear in the next few years since it's relatively easy to set one up and there are many ways to offer unique value.

As technology advances and businesses respond to our need to find information, we'll have richer and more satisfying ways to find information on the Internet. Technology and editorial services will combine to help us find information in a number of ways. The ideal navigation system will expose seemingly limitless amounts of information and yet remain easy to use.

Queries, filters, spatial navigation, links, and agents will be five of the primary selection techniques, as they are already to varying degrees. One way to understand the differences among these five approaches to finding information is to think of them metaphorically. Imagine specific information—a collection of facts, a breaking news story, a list of movies—all residing in an imaginary warehouse. A query does a search through every item in the warehouse to see if it meets some criterion you've set out. A filter is a check on everything new that comes into the warehouse to see if it matches that criterion. Spatial navigation is walking around inside the warehouse, checking on inventory by location. Links, whether text or graphical, take you directly to related information. Perhaps the most intriguing approach, and the one that promises to be the easiest of all to use, will be to enlist the aid of a personal agent who will represent you on the network. Most often the agent will be software, but it will have a personality you'll be able to talk to, or at least communicate with, in one way or another. Using an agent will be like delegating an assistant to look at the information inventory for you.

Here's how the different selection methods will work. With a query system, you'll be able to ask a wide range of questions and get complete answers. If you can't remember the name of a movie but you do remember that it starred Spencer Tracy and Katharine Hepburn and that there's a scene in which he's asking a lot of questions and she's shivering, you can type in a query that asks for all movies that match "Spencer Tracy," "Katharine Hepburn," "cold," and "questions." In reply, a server somewhere on the net will list the 1957 romantic comedy *Desk Set*, in which Tracy quizzes a shivering Hepburn on a rooftop terrace in the middle of the winter. You could watch the scene, watch the whole film, read the script, read the reviews, and read any public comments Tracy or Hepburn might have made about the scene. If dubbed or subtitled prints had been made for release outside English-speaking countries, you could watch the foreign versions. They might be stored on servers in various countries but would be instantly available to you.

The query system will accommodate straightforward requests, such as "Show me all the articles that ran worldwide about the first test-tube baby," or "List all the stores that carry two or more kinds of dog food and that will deliver a case within sixty minutes to my home address," or "Which of my relatives have I been out of touch with for more than three months?" It will also be able to deliver answers to much more complex queries. You might ask, "Which major city has the greatest percentage of the people who watch rock videos and regularly read about international trade?" Generally, queries won't require much response time because most of the questions are likely to have been asked before and the answers will already have been computed and stored.

Filters are really just standing queries. A filter works around the clock, watching for new information that matches an interest of yours and filtering out everything else. You'll be able to program a filter to gather information on your particular interests, such as news about local sports teams or particular kinds of scientific discoveries. If the most important thing to you is the weather, your filter will put that at the top of your personalized newspaper. Your computer will create some filters for you, basing the filters on information about your background and areas of interest. Such a filter might alert me to an important event that involves a person or an institution from my past: "Meteorite crashes into Lakeside School." You'll also be able to create explicit filters, on-going requests for something particular, such as "Wanted: 1990 Nissan Maxima for parts," or "Tell me about anybody selling memorabilia from the last World Cup final," or "Is anyone around here looking for someone to bicycle with on Sunday afternoons, rain or shine?" Such a filter will keep looking until you call off the search. If a filter does find a potential Sunday bicycling companion, it will automatically check on any other information the person might have published on the network, trying to answer the question, "What's he like?"—which is the first question you'd be likely to ask about a potential new friend.

Spatial navigation will be modeled on the physical way we locate information today. When we want to find out about some subject now, we may go to a labeled section of a library or a bookstore. Newspapers have sports, real estate, and business sections where people "go" for certain kinds of news. In most newspapers, weather reports appear in the same general location day after day.

Spatial navigation, which is already being used in some software products, will let you go where the information is by enabling you to interact with a visual model of a real or make-believe world. You can think of such a model as a map or an illustrated, three-dimensional table of contents. Spatial navigation will be particularly important for interacting with televisions and small, portable PCs, which are unlikely to have conventional keyboards. To do some banking, you might go to a drawing of a main street and then point, using a mouse or a remote control or even your finger, at the drawing of a bank. You might point to a courthouse to find out which cases are being heard by which judges or what the backlog is. You'd point to the bus terminal to find out what the schedule is and whether the bus you want to take is running on time. If you're considering staying at a particular hotel, you'll be able to find out when rooms are available and look at a floor plan. If the hotel has a video camera connected to the Internet, you might be able to look at its lobby and restaurant and see how crowded it is at the moment.

You'll be able to jump into any map so that you can navigate down a street or through the rooms of a building. You'll be able to zoom in and out and pan around to different locations. Let's say you want to buy a lawn mower. If the screen shows the inside of a house, you might go out the back door and see landmarks, including a garage. A click on the garage will take you inside it, where you might see tools, including a lawn mower. A click on the lawn mower will take you to categories of relevant information, including ads, reviews, user manuals, and sales showrooms in cyberspace. It will be simple to do some quick comparison shopping on the basis of any amount of information you want to assemble. When you click on the picture of the garage and seem to move inside it, behind-the-scenes information about the objects "inside" the garage will be fed to your screen from servers spread out over thousands of miles on the net.

When you point at an object on the screen to bring up information about it, you're using a form of "hyperlinking." Hyperlinks, or simply links, let users leap from informational place to informational place instantly, just as spaceships in science fiction jump from geographic place to geographic place through "hyperspace." Anyone who has used the Internet's World Wide Web is acquainted with links, which take the form of buttons, pictures, and underlined words that you click on to take you to another place in the document, another page on the Web site, or another site altogether. On the future broadband network, links will let you find answers to your questions as they occur to you, when you're interested. Let's say you're watching the news and you see somebody you don't recognize walking with the British prime minister. You want to know who she is. Using your television's remote control or your PC's mouse, you point at the person. That action brings up a biography and a list of other news accounts in which the person has figured recently. Point at something on the list, and you'll be able to read the news story or watch it, jumping any number of times from topic to topic and gathering video, audio, and text information from all over the world.

Spatial navigation can be used for touring too. If you want to see reproductions of the art in a museum or a gallery, you can "walk" through a visual representation of the place, navigating among the works much as if you were physically there. For details about a painting or a sculpture, you can use a link. No crowds, no rush, and you can ask anything without worrying about seeming uninformed. You'll bump into interesting things, just as you do in a real gallery. No, navigating through a virtual gallery isn't exactly like walking through a real art gallery, but it will be a gratifying approximation—just as watching a ballet or a basketball game on TV can be entertaining even though you're not in the theater or the field house.

If other people are visiting the same "museum," you can choose to see them and interact with them or not, as you please. Your visits don't have to be solitary experiences. Some locations on the net will be used purely for cyberspace socialization. In others nobody will be visible. Some locations will force you to appear to some degree as you are. Others won't. The way you look to other users will depend on your choices and on the rules of the particular location.

Museums all over the world already have Web sites on the Internet, although so far most of them display only a small number of images. At the best museum sites, such as the one for the Louvre in Paris (http://www.culture.fr/louvre), you can explore collections at length.

The browsing experience is surprisingly rewarding, although it's nothing like being in Paris.

If you're using spatial navigation, the place you're moving around in doesn't have to be real. Eventually you'll be able to set up imaginary places and return to them whenever you want to. In your own museum, you'll be able to move walls, add imaginary galleries, and rearrange the art. You might want all still lifes to be displayed together, even if one is a fragment of a Pompeian fresco that hangs in a gallery of ancient Roman art and one is a Cubist Picasso from a twentieth-century gallery. You'll be able to play curator and gather images of your favorite artworks from around the world to "hang" in a gallery of your own. Suppose you want to include a painting you remember with great affection of a man asleep being nuzzled by a lion. If you can't remember either the artist or where you saw it, you'll be able to describe what you want by posing a query. The query will start your computer or other information appliance sifting through reservoirs of information to deliver the piece that matches your request.

You'll be able to give your friends tours of your gallery, regardless of whether they're sitting next to you or watching from the other side of the world. "Here, between the Raphael and the Modigliani," you might say, "is a favorite finger painting I did when I was three years old."

The last type of navigational aid, and in many ways the most useful of all, is an agent—a query or a filter that has taken on a personality and that seems to show initiative. An agent's job is to help you. In the Information Age, of course, that means the agent is there to help you find information.

To understand the ways in which an agent can help with a variety of tasks, consider how an agent could improve today's PC interface. The present state of the art is the graphical user interface, as in the Apple Macintosh and Microsoft Windows operating systems. A graphical user interface depicts information and relationships on the screen graphically instead of just describing them in text. A graphical interface also allows the user to point to and move objects around on the screen.

But the graphical user interface has its limits. We've already put so many graphical options on the screen that programs or features that don't get used regularly have become daunting to users. The features are convenient for people who use them often enough to remember how they work, but for the average user the graphical user interface by itself doesn't provide enough guidance to using the features. In the systems of the future, agents will remedy that.

An agent will know how to help you partly because the computer will remember your past activities. It will note patterns of use that will help it work more effectively with you. Through the magic of software, information appliances connected to the net will seem to learn from your activities and will make suggestions to you. I call this "softer software."

The first day a new assistant is on the job, you can't simply ask him to format a document the way another memo you wrote a few weeks ago was formatted. You can't say, "Send a copy to everybody who should know about this." But over the course of months and years, the assistant becomes more valuable as he picks up on what is routine and on how you like things to be done.

The computer today is like a perpetual first-day assistant. It needs explicit first-day instructions all the time, remaining a first-day assistant forever. It will never make one iota of adjustment in response to its experience with you. At Microsoft, we're working to make software softer. No one should be stuck with an assistant who doesn't learn from experience.

Software enables hardware to perform a number of functions, but once a program is written, it usually stays the same. Softer software will seem to get smarter as you use it. It will learn about your requirements and preferences in much the way a human assistant does, and like a good human assistant, it will become more helpful the more it learns about you and your work.

If an agent that could learn were available now, I'd want it to take over certain tasks for me. For instance, it would be very helpful to me if it could scan every project schedule at Microsoft, note changes, and distinguish the ones I had to pay attention to from the ones I didn't. The agent would learn the criteria for what needed my attention: the size of the project, what other projects are dependent on it, the cause and the length of any delay. It would learn when a two-week slip in a schedule can be ignored and when such a slip indicates real trouble that I'd better look into right away, before it gets worse. Achieving this level of service

---------------------------->

will take the software industry time, partly because it's difficult, as with a human assistant, to find the right balance between initiative and routine. We don't want to overdo it. If a built-in agent tries to be too smart and anticipates and confidently performs services a user doesn't want, it will be annoying to users who are accustomed to having explicit control over their computers.

When you use an agent, you'll be in a dialogue with a program that behaves to some degree like a person. The software might mimic the behavior of a celebrity or a cartoon character as it helps you. An agent that takes on a personality provides a "social user interface." A number of companies, including Microsoft, are developing agents with social user interface capabilities. An agent won't replace graphical user interface software. It will supplement the graphical user interface by providing a character you choose to get to know you and help you. The character will disappear when you get to the parts of the software you know really well. But if you hesitate or ask for help, the agent will reappear and offer assistance. You may even come to think of the agent as a collaborator built right into the software. It will remember what you're good at and what you've done in the past and try to anticipate problems and suggest solutions. It will bring anything unusual to your attention. If you work on something for a few minutes and then decide to discard the revision, the agent might ask if you're sure you want to throw the work away. Some of today's software already does that. But if you were to work for two hours and then give an instruction to delete what you'd just done, the social interface would recognize that as unusual and possibly a serious mistake on your part. The agent would say, "You've worked on this for two hours. Are you really, really sure you want to delete it?"

Some people hearing about softer software and the social interface find the idea of a humanized computer creepy. But I believe even they will come to like it once they've tried it. We humans tend to anthropomorphize anyway. Animated movies take advantage of this tendency. *The Lion King* is not a very realistic depiction of animals in the jungle, nor does it try to be. We know very well that a real lion cub doesn't have the human characteristics Disney's animators attribute to Simba, but the fiction that Simba is like a human seems perfectly natural to us as we watch the movie. When a car breaks down or a computer crashes, we're apt to yell at it, or curse it, or even ask it why it let us down. We know better, of course, but we still tend to treat inanimate objects as if they were alive and had free will. Researchers at universities and software companies are exploring ways to use this human tendency to make computer interfaces more effective. Studies have shown that users' reactions differ depending on whether an agent's voice is female or male. People will treat mechanical agents that have personalities with a surprising degree of deference. We recently worked on a project in which we asked users to rate their experiences with a computer. When we had the computer the users had worked with ask for an evaluation of its performance, the people's responses tended to be positive. But when we had a second computer ask the same people to rate their encounters with the first machine, the people were significantly more critical. Their reluctance to criticize the first computer "to its face" suggests that they didn't want to hurt its feelings even though they knew full well that it was only a machine. Social interfaces may not be suitable for all users or all situations, but I think we'll see lots of them in the future because they "humanize" computers.

Agents will need to "know" about you in order to help you. With so much information about you and your preferences stored on the network, strong privacy and security measures will be essential. Fortunately, technology makes strong privacy measures feasible. In fact, privacy technology is so potent that it makes a lot of governments nervous.

Governments have long understood the importance of keeping information private, for both economic and military reasons. The need to make personal, commercial, military, or diplomatic messages secure (or to break into them) has attracted powerful intellects through the generations. Charles Babbage, who made dramatic advances in the art of code breaking in the mid-1800s, wrote, "Deciphering is, in my opinion, one of the most fascinating of arts, and I fear I have wasted upon it more time than it deserves." It's very satisfying to break an encoded message. I discovered this as a kid when, like kids everywhere, a bunch of us played with simple ciphers. We would encode messages by substituting one letter of the alphabet for another. If a friend sent me a cipher that began "ULFW NZXX," it would be fairly easy to guess that this represented "DEAR BILL" and that U stood for D, and L for E, and so on. With those seven letters it wasn't hard to unravel the rest of the cipher fairly quickly.

Wars have been won or lost because the most powerful governments on earth didn't have the cryptological power any interested junior high school student with a personal computer can harness today. Soon any child old enough to use a computer will be able to transmit encoded messages that no government on earth will find easy to decipher. This is one of the profound implications of the spread of fantastic computing power.

Encryption is shaping up as a key element in network communications, and for good reason. Without secure transactions, including financial transactions, it's hard to imagine that any interactive network will become a healthy marketplace.

When you send a message on the information highway, it will be "signed" by your computer or other information appliance with a digital signature that only you are capable of applying to it, and it will be encrypted so that only the intended recipient will be able to decipher it. You'll send a message, which could be information of any kind, including voice, video, or digital money. The recipient will be positive that the message is really from you, that it was sent at exactly the indicated time, that it has not been tampered with in the slightest, and that others can't decipher it.

The mechanism that will make this security possible is based on "one-way functions" and "public key encryption." I'm going to only touch on these advanced concepts. Keep in mind that regardless of how complicated the security system is technically, it will be extremely easy for you to use. You'll just tell your information appliance what you want it to do, and it will seem to happen effortlessly—and securely.

For encryption, we need a one-way function—a function that's much easier to do than to undo. Breaking a pane of glass is a one-way function, but not a function useful for encryption. For cryptography, the function must be easy to undo if you know an extra piece of information and very difficult to undo without that information. There are a number of such one-way functions in mathematics. One involves prime numbers. Kids learn about prime numbers in school. A prime number can't be divided evenly by any number except 1 and itself. Among the first dozen numbers, the primes are 2, 3, 5, 7, and 11. The numbers 4, 6, 8, 10, and 12 are not prime because 2 divides into each of them evenly. The number 9 is not prime because 3 divides into it evenly. There are an infinite number of prime numbers, and there is no known pattern to them except that they are prime. When you multiply two prime numbers together, you get a number that can be divided evenly only by those same two primes. For example, only the primes 5 and 7 can be divided evenly into 35. Finding the primes is called "factoring" the number. If we factor 35 we get the 5 and 7.

It's easy to multiply the prime numbers 11,927 and 20,903 and get the product 249,310,081, but it's much harder to recover from the product the two prime numbers that are its factors. This one-way function, multiplying prime numbers that are difficult to factor out of the product, underlies an ingenious and sophisticated encryption system. It takes a long time for even the biggest computers to factor a really large product back into its constituent primes. A coding system based on factoring uses two different decoding keys, one to encipher a message and a different but related one to decipher it. With only the enciphering key, it's easy to encode a message, but deciphering it within any practical period of time is nearly impossible. Deciphering requires a separate key, available only to the intended recipient of the message—or rather, to the recipient's computer. The enciphering key is based on the product of two huge prime numbers, whereas the deciphering key is based on the primes themselves. A computer can generate a new pair of unique keys in a flash because it's easy for a computer to generate two large prime numbers and multiply them together. The enciphering key thus created can be made public without appreciable risk because of the difficulty even another computer would have factoring it to obtain the deciphering key.

The practical application of this encryption function is at the center of various network security systems. You can think of future versions of the Internet as a postal network where everybody has a mailbox that is impervious to tampering, that has an unbreakable lock. Each mailbox has a slot that lets anyone slide information in, but only the owner of the mailbox has the key to get information out.

Each user's computer or other information appliance will use prime

------------------------------➤

numbers to generate an enciphering key, which will be listed publicly, and a corresponding deciphering key, which only the user will know. This is how it will work in practice: I have information I want to send you. My information appliance/computer system looks up your public key and uses it to encrypt the information before sending it. No one can read the message, even though your key is public knowledge, because your public key doesn't contain the information needed for decryption. You receive the message, and your computer decrypts it with a private key that corresponds to your public key.

You want to answer. Your computer looks up my public key and uses it to encrypt your reply. No one else can read the message, even though it was encrypted with a key that's totally public. Only I can read it because only I have the private deciphering key. This is a very practical encryption system because no one has to trade keys in advance.

How big do the prime numbers and their products have to be to ensure an effective one-way function?

Public key encryption was invented by Whitfield Diffie and Martin Hellman in 1977. Another group of computer scientists, Ron Rivest, Adi Shamir, and Leonard Adelman, soon came up with the notion of using prime factorization as part of what is now known as the RSA cryptosystem, after the initials of their last names. They projected that it would take millions of years to factor a 130-digit number that was the product of two primes, regardless of how much computing power was brought to bear on finding the solution. To prove the point, they challenged the world to find the two factors in this 129-digit number, known to people in the field as RSA 129:

114,381,625,757,888,867,669,235,779,976,146,612,010,218,296,
721,242,362,562,561,842,935,706,935,245,733,897,830,597,123,
563,958,705,058,989,075,147,599,290,026,879,543,541

They were sure that a message they had encrypted using this number as the public key would be totally secure forever. But they hadn't anticipated either the full effects of Moore's Law, as discussed in chapter 2, which has made computers much more powerful, or the success of the personal computer, which has dramatically increased the number of computers and computer users in the world. In 1993 a group of more than 600 academics and hobbyists from around the world began a methodical assault on the 129-digit number, using the Internet to coordinate the work of various computers. In less than a year they factored the number into two primes, one 64 digits long and the other 65:

3,490,529,510,847,650,949,147,849,619,903,898,133,417,764,638,
493,387,843,990,820,577

and

32,769,132,993,266,709,549,961,988,190,834,461,413,177,642,
967,992,942,539,798,288,533

The encoded message says, "The magic words are squeamish and ossifrage."

One lesson that came out of this challenge is that a 129-digit public key is not long enough if the information being encrypted is really important and sensitive. Another is that no one should get too cocksure about the security of encryption.

Increasing the key just a few digits makes it much more difficult to crack. Mathematicians today believe that a 250-digit-long product of two primes would take millions of years to factor with any foreseeable amount of future computing power. But who really knows? This uncertainty—and the unlikely but conceivable possibility that someone could come up with an easy way of factoring big numbers—means that networks will have to be designed in ways that allow encryption schemes to be changed readily.

The importance of this need for flexibility was demonstrated late in 1995 when a cryptography consultant, Paul Kocher, was able to break certain implementations of public key encryption without brute-force factoring. Kocher used the public key of a target system to encrypt messages and then carefully timed how long it took the system to decrypt them. After a few thousand tries at most, he had enough timing information to ascertain the secret number the system used for decryption. Fortunately, it's easy to modify decryption software so that its work can't be accurately timed. Unfortunately, there may be other ways of breaking public key encryption that nobody's thought of yet. So there's always something to worry about.

One thing we don't have to worry about is running out of prime numbers, or the prospect of two computers' accidentally using the same numbers as keys. There are far more prime numbers of appropriate length than there are atoms in the universe, so the chance of an accidental duplication is vanishingly small.

Key encryption allows more than just privacy. It can also assure the recipient of the authenticity of a document because a private key can be used to encode a message that only the public key can decode. It works like this: If I have information I want to sign before sending it to you, my computer uses my private key to encipher it. Now the message can be read only if my public key—which you and everyone else know—is used to decipher it. This message is verifiably from me because no one else has the private key that could have encrypted it in this way.

My computer takes this enciphered message and enciphers it again, this time using your public key. Then it sends this double-coded message to you across the information highway.

Your computer receives the message and uses your private key to decipher it. This removes the second level of encoding but leaves the level I applied with my private key. Then your computer uses my public key to decipher the message again. Because it really is from me, the message deciphers correctly and you know it is authentic. If even one bit of information was changed, the message would not decode properly and the tampering or communications error would be apparent. This extraordinary degree of security will enable you to transact business with strangers or even people you distrust because you'll be able to be sure that digital money is valid and that signatures and documents are provably authentic.

Security can be increased further by having time stamps incorporated into encrypted messages. If anyone tries to tinker with the time a document indicates it was written or sent, the tinkering will be detectable. Time stamps will rehabilitate the evidentiary value of photographs and videos, which has been under assault because digital retouching has become so easy to do.

My description of public key encryption simplifies the details and the drawbacks of the system. For one thing, because it is relatively slow, public key encryption will not be the only form of enciphering used on the highway. But public key encryption will be the method for signing documents, establishing authenticity, and securely distributing the keys for other kinds of encryption systems.

Navigation capabilities, a user interface, and encryption facilities are examples of software components that are not applications in and of themselves. They're standard services that an operating system provides to all of the applications built on top of it. Not every application running on an interactive network will take advantage of all of the available services, and many applications may extend the operating system's basic services by adding new features. Nevertheless, foundation services are vital. They help users by promoting consistency in the way applications are used, and they help software developers and content publishers by doing a lot of the hard work, making it faster and easier to create great applications of all kinds.

And make no mistake, there will be great applications of all kinds on the Internet—much better and far more plentiful than the ones available today. Many of tomorrow's net applications will be purely for fun, as they are today. People already use the Internet and on-line services to play bridge and board games with friends in other cities, but the experience will get much better. Televised sports events will offer you the opportunity to choose the camera angles, the replays, and even the commentators for your version. You'll be able to listen to any song, anytime, anywhere, piped in from the world's largest record store: the net. You might hum a little tune of your own into a microphone and then play it back to hear what it could sound like if it were orchestrated or performed by a rock group. You might watch *Gone With the Wind* with your own face and voice replacing Vivien Leigh's (or Clark Gable's). Or see yourself walking down a runway wearing the latest Paris fashions adjusted to fit your body or the one you wish you had.

Users who are just naturally curious will be mesmerized by the abundance of information available to them. Want to know how a mechanical clock works? You'll peer inside one from any vantage point and be able to ask questions. Eventually you may even be able to crawl around inside the clock, using a virtual reality application. Or you'll be able to assume the role of a heart surgeon or play the drums at a sold-out rock concert, thanks to the net's ability to deliver rich simulations to home computers. Some of the opportunities you'll have on the interactive network will just be successors of the opportunities available today through software and Internet connections, but the graphics and animation will be far, far better.

Other applications will be strictly practical. When you go on vacation, for example, a home-management application will be able to turn down the heat, ask the post office and the newspaper carrier to hold your mail and newspaper deliveries, cycle your indoor lighting so that you seem to be at home, and automatically pay routine bills. The technology for many of these kinds of services exists already.

Still other applications will be completely serious. My dad broke his finger one weekend a few years ago and went to the nearest emergency room, which happened to be Children's Hospital in Seattle. They refused to do anything for him because he was a few decades too old. Had there been a fully realized interactive network at the time, it would have saved him some trouble by telling him not to bother trying that hospital. An application communicating on the net would have told him which nearby emergency rooms were in the best position to help him at that particular time.

If my dad were to break another finger a few years from now, he not only would be able to use a net application to find an appropriate hospital, but he might even be able to register electronically with the hospital while driving there and avoid the usual paperwork holdup entirely. The hospital's computer would match his injury to a suitable doctor, who would be able to retrieve my father's medical records from a server on the net. If the doctor called for X rays, they'd be stored in digital form on a server, available for immediate review by any authorized doctor or specialist throughout the hospital or the world. Comments made by anyone reviewing the X rays, whether oral or in text form, would be linked to Dad's medical records. Afterward my father would be able to look at the X rays from home and listen to the professional commentary. And he'd be able to show the X rays to the family: "Look at the size of that fracture! Listen to what the doctor said about it!"

Most of these kinds of applications, from checking a pizza menu to sharing centralized medical records, are already starting to appear on PCs linked to the Internet. Features get better all the time and bandwidth will only increase, so the on-line experience will continue to get richer. Interactive information sharing will become a part of everyday life.

5

FROM INTERNET
TO HIGHWAY

The information highway doesn't exist. That may come as a surprise
to people who've heard everything from a long-distance telephone
network to the Internet described as a data "superhighway." Although
the Internet is already delivering communication services and information to millions of people, a broadband interactive network—able to
deliver all the killer applications described in chapter 4—won't be available to the majority of U.S. homes for at least a decade. We simply won't
have the high-speed infrastructure in place before then.

The Internet is the precursor of the ultimate global network. There
is little doubt that when the global interactive network has finally
evolved into the highway, it will still be called the Internet. But as quaint
as the term "information highway" is beginning to sound, using it
appropriately helps to draw a distinction between today's primarily
narrowband interactive network (the current "Internet") and tomorrow's
broadband interactive network (the "highway").

Whatever it ends up being called, constructing an interactive network
of enormous capacity ("bandwidth") is a big job. It calls for a dramatic

evolution of the Internet's software and hardware platforms. The heated competition we saw in the PC software industry during the 1980s is under way again, this time to determine which software components will become standard on the evolving Internet. Creating the network also requires the installation of expensive physical infrastructure: lines, switches, and servers. The timetable for all of this development and investment may not be clear for some time. We do know that businesses will connect up rapidly but homes will come on-line more slowly—first graduating from narrowband to midband connections that will make the Internet more useful than it is today and then eventually moving up to the fiber-optic broadband connections that will deliver on the full promise of the information highway.

There are no precise boundaries between narrowband and midband and between midband and broadband data networks. As a practical matter, though, a narrowband connection allows the transfer of at most about 50,000 bits of information per second to and from a single information appliance while a broadband connection accommodates continuous transmissions of at least 2 million bits per second—and for even better video quality, 6 million bits per second or more. As its name suggests, a midband connection falls in between narrowband and broadband in the number of bits per second of information it can deliver on a sustained basis to a particular computer, TV, or other information appliance. This classification of bandwidths can be helpful, but it's somewhat oversimplified because some connections will be asymmetric, able to receive information far faster than they can send it. Other date pipelines will be shared, so the speed of your connection will depend on how many other users are competing to use the same line.

Today nearly all residential connections to interactive networks are narrowband. Most consumers who plug into on-line services or the Internet use the telephone network's conventional "twisted-pair" copper wires, a narrowband system that relies on analog tones to communicate information. A "modem" (shorthand for **mo**dulator-**dem**odulator) is a hardware device that connects a PC to a phone line, serving as a translator between the digital and analog worlds. Modems convert a computer's digital information (0s and 1s) into patterns of tones that telephone networks can carry, and vice versa. In the early days of the

IBM PC, a modem typically carried data at the rate of 300 or 1,200 bits per second, also known as 300 or 1,200 "baud." Just about the only information you could transmit was text because at those low speeds transmitting pictures was agonizingly slow. Now typical modems send and receive 28,800 (28.8K) bits per second. This is a less agonizing rate, but it's still not fast enough to gracefully accommodate many rich forms of content. You can send a page of text in a second, but a complete screen-size photograph, even compressed, can take tens of seconds of transmission time. It takes whole minutes to send a color photograph that has the resolution of a slide. And just forget about high-quality motion video.

There are modems capable of transmitting data at 33,600 baud or even 38,400 baud, but modems won't be able to get much faster than this using normal phone lines. This is one reason the world is moving away from narrowband analog networks, which were designed to carry voice, and toward digital networks that carry much greater amounts of information per second.

In the years ahead telephone and cable companies will upgrade their networks with new digital switches and with fiber-optic cable, which has far greater bandwidth than copper wire. Once the new infrastructure is in place, the era of broadband communications—the era of the information highway—will have arrived.

Digital switches aren't like light switches. They don't turn anything on or off. They're more like trainyard switches that route railcars from one track to another. Think of a stream of information, such as an e-mail message or a movie, as a "train" of bits that leaves one railway station (a server) and makes its way across the network to a destination railway station (your home PC or TV). The network has countless switches, each of which can route information toward its destination. You might expect that the job of the switches would be to find the shortest route for the train, but it isn't. Routing the whole train all at once would tie up the tracks and any station through which the train ran. It would be like a locomotive pulling a long line of boxcars: Nothing else would get through until the train had passed. To circumvent this problem, the network's digital switches break the train up into individual "boxcars" of digital information, and in some cases route each car independently.

Different parts of a single e-mail message, for example, may travel very different routes, each boxcar finding its own way through the switches to the destination—where the cars are reassembled into a coherent "train" of information. When you order a movie, though, all of the boxcars will follow a single route to assure smooth delivery, but even so the cars will have a lot of room between them so that other streams of content passing through the same switches can intermingle with them. The packages of information aren't called "boxcars," of course. They're called "packets" or "cells."

The routing of packets will most likely be accomplished through the use of a communications protocol known as asynchronous transfer mode, or ATM (not to be confused with "automatic teller machine"). Long-distance telephone companies around the world are enthusiastic about the ATM protocol. According to the rules of the protocol, a digital stream of data is broken into packets, each of which consists of 48 bytes of the information to be transported and 5 bytes of control information that enable a network's switches to route the packet to the destination. A second of video, for example, gets broken into thousands of packets, each of which is individually addressed and sent to the destination. ATM switches have the capacity to handle many different streams of information at once because the streams are delivered at very high speeds—up to 622 million bits per second with existing technology and eventually at 2 billion bits per second or more.

Communications lines need to have great capacity to handle this much information at once, and that's where fiber-optic cable comes into play. Fiber, the drawn glass that carries information in a beam of modulated light, will be the "asphalt" of the information highway—or the "tracks" of the information railway, to extend our analogy. All of the major long-distance trunk lines that carry telephone calls between "switchyards" in the United States already use fiber. Parts of some cable television networks do too. The network companies weren't thinking of ATM when they began installing fiber, but they're keenly conscious of it now. Telephone and cable companies will gradually run fiber-optic cable to neighborhood distribution points, where the signals will be transferred either to the coaxial cable that brings you cable television or to the twisted-pair, copper wire lines that give you telephone service. Eventually fiber may run directly into your home, especially if you use lots of data. But it's not necessary to have fiber running all the way to your home to get great broadband service.

Except in business districts and other areas where there is a high density of people willing to pay for the connections, broadband networks won't be widespread for a number of years. Everybody pretty much agrees now that the installation of the residential infrastructure is going to be a prolonged process. That realization reflects quite a change in thinking from the mindset that prevailed in 1994, when phone companies and cable companies were promising that millions of broadband residential connections would show up within a few years. It was a period remarkable for the extravagance of its promises. Several major U.S. telephone companies each proclaimed that they would wire 500,000 to 1 million households a year to broadband networks throughout the decade, with a total of more than 9 million expected to be hooked up by the end of 1996. Bell Atlantic said it would wire as many as 1.5 million households in some years, 8 million by the year 2000—although it acknowledged the fact that giving people access to a high-speed connection didn't necessarily mean that they'd order broadband services. Pacific Bell said 5.5 million of its customers would have broadband connections by the end of the decade, and Ameritech said it would have 6 million of its households hooked up. Microsoft was considered conservative because our reaction to these claims was cautious: We said that the new infrastructure shouldn't be built until we had a good trial that showed there would be enough revenue to justify the large investment.

The realization that broadband delivery won't happen overnight came as companies recognized that the costs of building a broadband system were higher than they expected and that they had very few applications for it. Just as important, the dramatic success of the Internet beginning in 1995 demonstrated that millions of people would use narrowband interactive networks enthusiastically for PC tasks. As I said in the preface, the sudden popularity of the Internet surprised me. My mistakes were underestimating how many people were willing to use a relatively slow network and underestimating the frenzy of Internet content generation that would take place. The Internet was fraught with problems, from the difficulty of setting up connections to a lack of privacy and security. By late 1995, though, it was clear that the Internet was achieving critical mass, generating a positive spiral greater than even the original PC had. Now every new Internet user makes it that much more worthwhile to create new content, and more content in turn attracts more users. The Internet's abrupt success is impressive testimony to the compelling nature of interactive networks—and a watershed event that has transformed everybody's expectations for the future.

When the term "information highway" came into vogue in the spring of 1993, many companies assumed that the communications revolution would take off with interactive television. Everybody thought that video applications would be the springboard from which digital networks would find commercial success. The TV set looked like the obvious device for delivering initial services because TVs were cheaper, more common, and easier to use than PCs. Most of the proposals for bringing interactivity into households focused on delivering video— not surprising since video-on-demand draws on the huge numbers of movies and TV shows already available and is a relatively easy application to implement. Cable and telephone companies rushed to get interactive TV trials going, both to learn about the market and to demonstrate to the world that they were poised on the cutting edge of The Next Big Thing. Some of these trials turned out to be pretty bogus from a technology standpoint. In one, for example, whenever a household ordered a movie what went on behind the scenes was almost comical: A videocassette was loaded by hand into a VCR in the cable company's office and played for the benefit of that one customer across town. This may have produced good marketing information, but it was not the information highway.

At Microsoft we agreed with the conventional wisdom that video-on-demand was a necessary killer application for broadband connections. We believed that a wide range of applications would have to be available from the very beginning if the network were to succeed—and we set out both to create a software platform that would support myriad applications and to create applications ourselves. In other words, we tried to do what we had done with Windows: to create a software platform and a collection of key applications that would help give the platform value. We didn't rush into trials, which prompted some journalists to say we were behind in the race to build the highway. In reality, we were preoccupied. We were busy fashioning a comprehensive architecture to support a full-featured network. We considered most of the early trials to be public relations exercises, distractions rather than opportunities, and we didn't want to get into a trial until we had something real to test.

Some companies didn't see PC software playing a role in the interactive market. We pushed for a strong compatibility and synergy between the PC and the TV, and we believed that corporate computer networks and the larger interactive network would converge over time, leading to a common architecture for the exchange of digital information.

Meanwhile the industry as a whole began to struggle with justifying the investments that would be necessary to bring the broadband network to tens of millions of households in a matter of only a few years. The cost of connecting a TV or a PC in a U.S. home to a broadband network was estimated at about $1,200, give or take a couple of hundred dollars. That price included running the fiber into the neighborhood, the servers, the switches, and the electronics in the home. With roughly 100 million homes in the United States, this worked out to around $120 billion for one country alone. Companies that had once planned aggressive investments began to worry about whether people would pay much for interactive services. They realized that new services would need to generate almost as much revenue again as cable television does and that entertainment applications alone wouldn't be enough. Marketing tests suggested that video-on-demand might generate only a little more revenue than simpler, cheaper systems that merely showed popular movies frequently. This news made it all the harder to justify building expensive new infrastructure on an accelerated basis.

I thought the solution would come from innovative new applications. When we met to discuss our highway efforts, I'd ask: "What are we doing on medical? What are we doing on travel? What are we doing on education? Hey, are we kidding ourselves here? Do these applications build excitement? How are we going to get all the stuff we need developed?" We knew that we would have to get lots of other companies building these applications too. However, without the network, it was difficult to get other companies to invest.

Once the industry realized that there would not be enough near-term revenue to justify the expensive broadband connections that had been promised, most of the forty-odd scheduled trials of interactive technology were canceled or scaled way back. Among the two-way broadband trials deployed were Time-Warner's in Orlando, Florida; Bell Atlantic's near Washington, D.C.; and one by the Japanese telecommunications giant, NTT, in Yokosuka, Japan. Microsoft and NEC were partners in the Japanese trial.

If a broadband network would be too expensive, what about midband as an evolutionary step? Even though this would eliminate video-on-demand as an application, technologies such as ISDN (integrated services digital network) and ADSL (asymmetrical digital subscriber line) could deliver midband data services (with only so-so video) over existing wires, saving a lot of money. But still the industry wasn't sure that a public who were assumed to want video first and foremost would settle for network connections that provided substandard video. Even though the costs were lower, we still needed applications. The interactive revolution seemed to be stalled.

Then, almost overnight, the Internet answered the questions that hung over the industry. It became clear that the interactive network would be built first around the personal computer and later around the TV, which would itself become more like a PC. There would be synergy between the PC, with its well-established authoring and content environments, and the eventual interactive TV environment. It also became clear that the public network (the so-called highway) and corporate networks would be similar and that they would interoperate with each other. And finally the industry realized that people would pay something for connectivity. It became apparent that early interactive content would not center on video-rich entertainment as much as on information and commerce applications—the kinds of content that work on narrowband networks but work even better on midband. Suddenly midband connections had a huge future and broadband service aimed at delivering video entertainment was off somewhere on the horizon.

The Internet is a loose collection of interconnecting commercial and noncommercial computer networks. The constituent networks are tied together by telecommunications lines and by their shared reliance on standard communications protocols (rules). This decentralized structure makes sense when you consider the origin of the Internet.

The Internet is an outgrowth of a government network called the ARPANET, which was created in 1969 by the Defense Department so that defense contractors and researchers could continue to communicate even after a nuclear attack. Rather than try to harden the network against nuclear weapons, ARPANET's designers decided to make it resilient by distributing its resources in a completely decentralized way—so that destruction of any part of the network, or even of most of it, wouldn't stop the overall flow of information. The network quickly found favor among computer scientists and engineers in industry and universities, and it became a vital communications link among far-flung collaborators. It was virtually unknown to outsiders.

Because the Internet originated in a computer science rather than a commercial environment, it has always been a magnet for hackers, some of whom have used their talents for breaking into computer systems. As a matter of fact, it was a hacker who first made many people in the U.S. aware that the Internet even existed. On November 2, 1988, thousands of computers connected to the Internet began to slow down. Many eventually ground to a temporary halt. No data were destroyed, but millions of dollars' worth of computing time was lost as computer system administrators fought to regain control of their machines. The cause turned out to be a mischievous computer program called a "worm" that was spreading from one computer to another on the network, replicating as it went. (It was designated a worm rather than a virus because it didn't infect other programs.) The worm used an unnoticed "back door" in the systems' software to directly access the memory of the computers it was attacking. There it hid itself and passed around misleading information that made it harder to detect and counteract. Within a few days the *New York Times* identified the hacker as Robert Morris, Jr., a twenty-three-year-old graduate student at Cornell University. Morris later testified that he had designed and then unleashed the worm to see how many computers it would reach but that a mistake in his programming had caused the worm to replicate far faster than he had expected. Morris was convicted of violating the 1986 Computer Fraud and Abuse Act, a federal offense. He was sentenced to three years of probation, a fine of $10,000, and 400 hours of community service.

----------------------------->

"On the Internet, nobody knows you're a dog."

In 1989, when the U.S. government elected to stop funding the ARPANET, users who depended on the network laid plans for a successor, to be called the "Internet." The word came to mean both the network itself and the protocols that governed communication across the network—a dual meaning that has been a source of confusion ever since. Even when it became a commercial service, the Internet's first customers were mostly research organizations, computer companies, university scientists, and graduate students, who used it to exchange e-mail.

It wasn't until recent years that the Internet became the backbone connecting all the world's different electronic mail systems. Today anybody can send anybody else a message on the Internet—for business, education, or just the fun of it. Students around the world send messages to each other. Shut-ins carry on an animated social life with friends they might never get out to meet. Correspondents who might be uncomfortable talking to each other in person are getting to know each other across the network. As bandwidth increases and computer processors get faster, network communications will include more video, which—unfortunately or not—will do away with the social, racial, and gender blindness that text-only exchanges permit.

A whole vocabulary and culture have grown up around e-mail. Some people attack other people deliberately, a practice known as "flaming," but users of e-mail in the early days sometimes found out the hard way that a humorous remark could be mistaken for a "flame." "Emoticons," little variations on print conventions, were developed to make the e-mail writer's humorous intentions unmistakable. If you want a sentence to end with a chuckle to show that you mean it to be humorous, you might add a colon, a dash, and a parenthesis. This composite symbol, :-), if viewed sideways, makes a smiling face. You might write, "I'm not sure that's a great idea :-)"—the smiling face showing that your words are meant in a good-natured way. Using the opposite parenthesis turns the smiling face into a frowning face, :-(, a comical expression of disappointment. Like social, racial, and gender blindness, these half cousins of the exclamation point may not survive the transition of e-mail into a medium that makes it convenient to incorporate audio and video in messages.

When I send you an e-mail message, it's transmitted from my computer to the server that has my "mailbox," and from there it passes directly or indirectly to whichever server stores your mailbox. When you connect to your server, via the telephone network or a corporate computer network, you can retrieve ("download") the contents of your mailbox, including my message. You can type a message once and send it to one person or twenty-five, or you can post it on a "bulletin board."

The cork bulletin board's electronic namesake is where messages are posted for anyone to read. Public conversations result as people respond to messages. These exchanges are usually asynchronous. Bulletin boards are generally organized by topic to serve specific communities of interest. This makes them an effective way to reach targeted groups. Commercial services offer bulletin boards for pilots, journalists, teachers, and many other smaller communities. On the Internet, the often unedited and unmoderated bulletin boards called "USENET newsgroups" have been around for twenty years. Thousands of communities are devoted to topics as narrow as caffeine (alt.drugs.caffeine), Ronald Reagan (alt.fan.ronald-reagan), and urban folklore (alt.folklore.urban). You can download all of the messages on a topic, or just recent messages, or all messages from a certain person, or all those that respond to a particular other message, or those that contain a specific word in their subject line, and so on.

Today the Internet's most popular application is "Web browsing." The "World Wide Web" is a network of servers connected to the Internet that offer pages of information containing text, graphics, sounds, and programming. When you connect to a Web server, an initial screen (or page) of information containing a number of links appears. When you activate a link by clicking on it with your mouse, you are taken to another page containing additional information and possibly

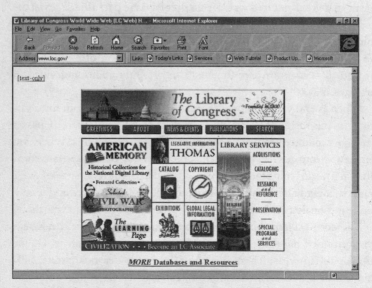

1996: U.S. Library of Congress home page on the World Wide Web, showing links

other links. That page may be stored on the same server as the first or on any other server connected to the Internet.

The main page at a company's or an individual's Internet "Web site" is called the "home page." If you create a home page, Internet users can find it by typing in the address (also called the "Uniform Resource Locator," or just "URL") or by clicking on a link to the address. In print and TV advertisements today we're starting to see a home page address in addition to a company's phone number. A Boeing advertisement, for instance, might include the address "http://www.boeing.com," Boeing's URL.

The software to set up a Web site is very cheap, and it's available for almost all computers. Software to browse the Web is also available for free for virtually all machines. You can Web browse using software on the CD that comes with some English-language editions of this book. PC operating systems are starting to integrate Web browsing, and many PC applications now enable users to both read and write Web pages.

Now that the Internet is benefiting from a positive-feedback cycle, it has established itself as a major distribution channel as well as an exciting forum for innovation. Many early predictions about interactive books and "hypertext"—made decades ago by pioneers such as Vannevar Bush and Ted Nelson—are coming true on the Web.

Literally every day now there is some new and interesting Internet development. The pace of the Internet's evolution is so fast that a description of the Internet as it existed a year or even six months ago might be seriously out-of-date now. It's hard to stay current with something so dynamic. Much of the mid-1990s Internet culture will seem as quaint in retrospect as stories of wagon trains and pioneers on the Oregon Trail do to us today.

In keeping with the highway metaphor, the Oregon Trail might even be a good analogy for the Internet. Between 1841 and the early 1860s, more than 300,000 pioneers rode wagon trains out of Independence, Missouri, for a dangerous 2,000-mile journey across a wilderness to the Oregon Territory or the gold fields of California. An estimated 20,000 people succumbed to marauders, cholera, starvation, or exposure. You could easily say that the route they took, the Oregon Trail, was the start of today's interstate highway system. It crossed many boundaries and provided for two-way vehicular traffic. The modern path of Interstate 84 and several other highways follows the pioneers' trail for much of its length. But the comparison between the Oregon Trail and the interstate does break down. Cholera and starvation aren't a problem on Interstate 84. And tailgating and drunk drivers weren't much of a hazard for wagon trains—although drunk cowboys might have been.

The trail blazed by the Internet leads toward the information highway. The early Internet has been a precursor of the broadband interactive network in much the same way that the Altair 8800 was a precursor of the modern personal computer. The Altair, that little computer Paul Allen and I wrote the first general-purpose microcomputer software for, was slow, primitive, and very difficult to use, but to a small but enthusiastic set of people it had seductive possibilities. Compared to the broadband interactive network we'll see in a few years, the Internet has been slow, primitive, and difficult to use. But it's improving even faster than the PC did, and it promises to be as important. Already the Internet's surging popularity is the most important single development in the world of computing since the IBM PC was introduced in 1981. The comparison with the PC is apt for other reasons too. The PC wasn't perfect. Aspects of it were arbitrary or even inferior technologically. Despite its shortcomings, its popularity grew to the point that the PC became the standard hardware for which applications were developed. Companies that tried to fight the PC standard often had good reasons, but their efforts failed because so many other companies were invested in continuing to improve the PC.

The lessons of the PC industry are not lost on companies such as Microsoft and Netscape, who are working together—and in competition—to define standards that will help overcome the Internet's limitations. Netscape and Microsoft are two of the companies supplying components of the software platform for interactive networks. We're all racing to embrace existing standards and extend them.

The success of the Internet has focused everyone's efforts on improving its standards. In 1994 Microsoft was going to build its own software platform for interactive networks, AT&T was going to build its own platform, IBM was going to build yet another platform, and Oracle was going to build one too. Each of these platforms was going to be a complete solution, tuned to the strengths and interests of the company that created it. Network operators—telephone and cable television companies, for instance—would choose a platform for their customers. The industry would figure out ways to let bits move around between the different systems, creating a limited degree of compatibility.

But now it is clear that there will be only one primary software platform, an evolutionary descendant of today's Internet. This platform will be the standard not only in the public networks around the world, but in corporate networks too—the "intranets" that serve employees but restrict the access of anyone outside the corporate network's electronic "firewall." The primary competition is no longer between different software platforms but is rather a competition to advance the evolution of the Internet. This is a significant change, and it motivated a lot of the changes between the first and second editions of this book.

The revolutionary approach won't vanish entirely. At least a few broadband networks that won't be supersets of the Internet are likely to be developed. Some governments, including Hong Kong's and Singapore's, may mandate and subsidize independent initiatives, for example. These efforts will be helpful because they'll encourage experimentation with applications and interfaces. But most of the effort and most of the progress will be built on the Internet platform.

The Internet's software foundation is strong. The TCP/IP protocol that governs transmissions supports distributed computing and scales very well. The protocols that define Web browsing are simple and have allowed servers to handle large amounts of traffic. But despite these good foundations, a great deal of work remains before the software platform will be complete. The platform must evolve to the point that it offers first-rate navigation and security, group collaboration capabilities, billing and accounting services, and connections to arbitrary software components. The platform must define standards for expressing user preferences and passing them between applications seamlessly.

To achieve its full potential, the Internet needs the ability to reserve bandwidth with a quality of service guarantee. Since ATM supports quality of service guarantees in its design, it will most likely be used as the underlying technology—although there are quality of service protocols that do not require ATM. With any of these protocols, a user can reserve bandwidth between two points on the network and get smooth delivery of real-time content such as audio and video. Quality of service guarantees are a necessary development if applications such as video-conferencing and video-on-demand are to succeed, but they will rob the Internet of a little more of its fading innocence by introducing payments into its traffic handling. In all likelihood, people will have to pay a small amount for any communication that has a quality of service guarantee. Without some form of differential pricing, everybody would request a quality of service guarantee for every communication. The Internet can't reserve bandwidth for all messages—that just won't work—so there has to be some way to charge users extra for the guaranteed performance. Right now the Internet is like a restaurant that doesn't accept reservations. If you want service, you just show up and accept whatever's available, confident that everybody else is in the same situation and that you're not at a disadvantage when it comes to getting prompt service. But because tomorrow's Internet will accept paid reservations, the only service you'll get without a reservation is whatever the unreserved bandwidth can accommodate. Whenever you use a voice connection, you will use a quality of service connection. The extra monthly cost for heavy voice use will be very small because audio connections don't require much bandwidth.

It costs remarkably little to use the Internet, except in countries where its use is subject to heavy surcharges or there aren't enough competing vendors to keep prices down. For what is often a flat monthly rate, the Internet provides U.S. customers with worldwide connections between servers, facilitating the exchange of electronic mail, bulletin board items, and other data. The exchanges range from short messages of a few dozen characters to multimillion-byte transfers of photographs, software, and other kinds of data. It costs no more for a user to request data from a server thousands of miles away than it does to ask for data from a server a mere mile away. The Internet's inexpensive pricing model has already done serious damage to the notion that communication has to be paid for by time and distance—just as inexpensive PC computing did away with hourly rates for computer time-sharing.

The financial model that enables the Internet to be almost suspiciously cheap is an extension of an interesting precedent that has long been in limited use. When you use a telephone today, you expect to be charged for time and distance unless you're phoning from inside a free local calling area. But businesses that call one remote site a great deal can avoid these time and distance charges by leasing a special-purpose telephone line dedicated to calls between the two sites. There are no traffic charges on a leased line—the monthly rate is the same no matter how much it's used.

The foundation of the Internet is a bunch of these leased lines connected by switching systems that route data. The long-distance Internet connections are provided in the United States by five companies who lease lines from telecommunications carriers. Since the breakup of AT&T in 1984, charges for leased lines have become quite competitive. Because the volume of traffic on the Internet is so large, the five companies qualify for the lowest possible rates—which means that they carry enormous amounts of data very inexpensively. They pass on their low rates in the form of low, usually flat rates to the companies that connect up to the Internet.

The rate is usually based on the capacity of the connection. A relatively high bandwidth connection called a "T-1 line" might cost $1,600 per month. The fee covers all of a company's Internet traffic, whether constant or rare and whether it goes a few miles or around the globe. This pricing structure means that once a customer has an Internet connection there is no extra cost for extensive use, which of course encourages use. The sum of the fees that users pay underwrites the entire Internet, without government subsidy.

Most individuals can't afford to lease a T-1 line. To connect to the Internet, they contact a local on-line service provider. This is a company that has paid to connect via T-1 or other high-speed means to the Internet. Individuals use their regular (narrowband) phone lines to call the local service provider, and it connects them to the Internet.

In 1994 and 1995 the typical monthly charge for Internet use in the United States was $20, for which a customer might get twenty hours of prime-time usage. In the spring of 1996 AT&T was among some companies that began offering totally unlimited usage for the same $20 price, signaling the arrival of a very competitive market for consumer access to the Internet. Other large telephone companies around the world are entering the Internet-access business too. On-line service companies such as CompuServe and America Online now include Internet access in their charges. As a truly mass market develops over the next few years, the Internet will continue to improve, benefiting from easy access, wide availability, easy navigation, and integration with commercial on-line services.

All of this puts the telephone companies in an interesting position. On the one hand, they may become the biggest providers of Internet access. On the other, the Internet threatens to take away much of the lucrative long-distance business that supports the telephone companies today. It's becoming more common for people to use the Internet for long-distance calls to other Internet users anywhere in the world—despite the poor quality of voice transmission. But as quality of service guarantees are incorporated into the Internet platform, the fidelity of both audio and video two-way calling will become quite impressive.

This will be great for consumers, but not all phone companies welcome the competition. In March of 1996 a trade group representing more than 130 small and medium-size telephone companies petitioned the United States Federal Communications Commission (FCC) to regulate and tax the sale of software and hardware products that enable the Internet to be used for long-distance services. A press release from the group, the America's Carriers Telecommunication Association, charged that the Internet "creates the ability to by-pass local, long-distance and international carriers and allows for calls to be made for virtually no cost." The statement warned that this traffic could overwhelm the Internet.

I don't blame the telephone companies for being upset. They're in a difficult position. Through the rate-making process, the U.S. government has put them in the position of losing money on some of the local-access service they provide. There have been good historical reasons for telephone rate regulation, such as the desire to make telephone service affordable to everyone—the "universal service" doctrine. Up until now U.S. phone companies have done fine by making up any losses with high-margin fees for long-distance service. In effect, overpriced long-distance service has subsidized underpriced local service. But to the degree that long-distance phone calls are replaced by the new form of communication on the Internet, the telephone companies will lose the subsidy for the loss-making local service. This is a serious matter, but it would be a big mistake to regulate Internet communications as if it were telephone communications, and I can't imagine its happening. For one thing, it would be difficult to define and detect phone calls among the packets of digital data that people exchange.

The phone companies are right when they say that the Internet could be overwhelmed by traffic from time to time as it grows. But this will be a passing problem because the ongoing investments in the Internet's infrastructure are heavy and because the bandwidth provided by communications technology is increasing rapidly. Demand will grow fast, but capacity will grow faster. The balancing act will be tricky, and there could be periods when the Internet's response times will be slow, but there's no fundamental roadblock that's going to stop the Internet—including its voice traffic—from flourishing.

If congestion becomes a major problem, one solution is to make everybody pay a higher flat rate. Another approach is to find something to meter—whether time on the system, the distance over which bits are transmitted, the number of bits, or whatever. It's not a straightforward situation, though, because of the large numbers of companies offering connections to the Internet. If only some of them have pricing schemes that hurt heavy users, the heavy users will migrate to companies that don't penalize them and the total amount of Internet traffic will be affected little if at all. (This is a phenomenon called "adverse selection" that will be familiar to anyone who works in the insurance industry: The least desirable potential customers tend to be the ones mostly likely to sign up because they have relatively more to gain.)

As the Internet is changing the way we pay for communication, it may also change how we pay for information—or don't pay for it. Most of the investment in Web-based publishing so far has been a labor of love, or an effort to help promote products sold in the nonelectronic world. Few content providers have been paid much directly by consumers, although many of them are looking forward to the day when they figure out how to get paid. A large interactive content industry has sprung up in which almost nobody makes any money so far.

Some prognosticators think that the Internet has shown that

information will be free, or largely so. It is true that a great deal of information, from NASA photos to bulletin board entries donated by users, will continue to be free. But I believe that much of the most attractive information, whether Hollywood movies or encyclopedic databases, will be produced with profit in mind.

It's hard to imagine the Internet thriving as a publishing medium unless content providers are paid for their work. There will be a lot of disappointment in the short run among content companies struggling to make ends meet through subscriptions or advertising, though. Advertisers usually hesitate to move into a new medium, and the Internet is certainly that. Some content companies are experimenting with a combination of subscriptions and advertising, but far fewer people will pay for content than will use it for free, so the subscriber base a content company can offer advertisers is reduced, which lowers ad rates and thus the content provider's ad revenue.

Another reason that charging for content doesn't work very well yet is that it's not practical to charge small amounts—or to pay small amounts. It isn't feasible to charge or pay 3 cents to read a news article. This temporary awkwardness will disappear as the Internet evolves. If you decide to visit a Web page that costs a dime, you'll pay the fee as part of a larger bill—just as you pay for all of your telephone service today on a monthly basis. I think we'll see a great deal of content offered at very low prices. After all, even 3-cent properties can make money if enough people visit them.

In addition to free information, there's a lot of free software on the Internet today, some of it quite useful. Sometimes it's commercial software given away as part of a marketing campaign. Other times the software has been written as a graduate student project or at a government-funded lab. But I think consumer desire for quality, support, and comprehensiveness in important software applications means that demand for commercial software will continue to grow. Already many of the students and faculty members who wrote free software at their universities are busy writing business plans for start-up companies that will provide commercial versions of their software with more features, not to mention customer support and maintenance.

One of the hottest areas of development, two-way wireless connections to the Internet, is destined to remain primarily narrowband for the foreseeable future. Police and medical personnel may buy enough bandwidth to make mobile two-way video practical, but for most people bandwidth will be considerably more limited. Satellites are already delivering streams of data to PCs, but the flow of data is primarily one-way. The wireless networks for mobile communication will grow out of today's cellular telephone systems and the alternative wireless phone service, called PCS. When you're on the road and want information from your home or office computer, your portable information appliance will connect to the wireless part of the highway, and a switch will transfer the connection to the wired part, which will connect the appliance to the computer/server in your home or office and bring you the information you asked for.

There will also be local, less expensive kinds of wireless networks available inside businesses and most homes. These networks will enable you to connect to the highway or your own computer system without paying time charges so long as you're within a certain geographical range. Local wireless networks will use technology different from that used by the wide-area wireless networks. However, a portable information device will automatically select the least expensive wireless network it's able to connect to, so as a user you won't be aware of the technological distinctions.

Another important technology that will work on a narrowband network is the sharing of a screen or a window on a screen so that it can be viewed by two or more people at different locations. Preliminary versions of this kind of capability are being built into Web browsers already, and some modems support the simultaneous transmission of voice and data across a single phone line. If you're making travel plans and you and your travel agent both have PCs equipped with an appropriate modem, she might show you photos of each of the different hotels you're considering, or display a grid comparing prices. If you call a friend to ask how to get to her house and you both have PCs connected to your phone lines, then during the conversation she'll be able to transmit a map to you, which the two of you can discuss or even annotate interactively.

A midband connection gives you about a factor of eight improvement, on average, compared to the narrow bandwidth of a conventional modem. Pictures come onto a screen very fast when you have a midband connection, and you can even start to use low-quality video, especially if it's in a small window on your PC screen. The additional bandwidth makes quite a difference in the experience of using the Web.

Telephone companies are already delivering midband connections into homes by feeding them digital rather than analog signals. Conventional wiring is used, but the switches that route information differ from those used in what the industry calls POTS, "plain old telephone service."

The primary phone company approach is to use ISDN (again, for integrated services digital network), which transfers voice and data starting at 64,000 or 128,000 bits per second. Motion video can be transmitted over ISDN lines, but the quality is mediocre at best—certainly not good enough for watching a movie and barely satisfactory for routine videoconferencing, although videoconferencing quality improves if multiple ISDN circuits are used simultaneously.

Hundreds of Microsoft employees use ISDN every day to connect their home computers to our corporate network. ISDN was invented more than a decade ago, but without the demand generated by PC applications, almost nobody needed it. It's remarkable that the phone companies invested enormous sums in switches to handle ISDN with very little idea of how it would be used. The good news is that the PC will finally drive explosive demand for ISDN. An add-in card for a PC to support ISDN once cost $500, but the price should drop to less than $200 over the next few years. The line costs vary by location but are generally about $50 per month in the United States. I expect that this rate will drop to less than $20, not much more than the cost of a regular phone connection.

One drawback of ISDN from a phone company's point of view is that an ISDN line ties up resources in the conventional voice network. Another midband technology, a new one that will grow in importance throughout the remainder of the decade, doesn't have this disadvantage. ADSL (again, for asymmetrical digital subscriber line) bypasses the voice network entirely. It uses conventional telephone wire to connect a PC with a telephone company's central office, where it is routed to the Internet.

But bypassing the voice network isn't the real selling point of ADSL. What makes it exciting is that it offers considerably more bandwidth into the home—although not always back out of the home—than ISDN does. ADSL can deliver at least 1.5 million bits per second into the home, the same as the T-1 lines that link many businesses to the Internet. Unfortunately, the data rate back out of the home (the "back channel" rate) can be as low as 64K bits per second—a lot faster than the rate a conventional modem offers, but not fast enough for a pleasant experience with two-way video. Fortunately, ADSL technology is improving, and I expect that many homes will be able to send data out to the network at up to 600K bits per second, a fast enough back channel rate for reasonable videoconferencing quality.

ADSL will also deliver higher speeds into the home, up to 6.2 million bits per second and sometimes more. But it can do this only when the distance between the telephone's central office and the PC is less than two miles.

ADSL was designed a few years ago as the telephone companies' answer to the threat of interactive TV, which the rival cable companies had promised to deliver. ADSL will be an excellent way to access the Internet. Web pages rich in graphics will appear on the screen almost instantly. The price of ADSL modems should drop to less than $300 as they become popular.

Telephone companies aren't the only players with the motivation and technology to put customers on the Internet at higher speeds. Cable companies intend to use their existing coaxial cable networks to connect PCs to the Internet as well as compete with the phone companies to provide local telephone service.

Coaxial cables, the kind that hook a TV to a VCR, have much higher bandwidth potential than standard telephone wires. Much of this enormous potential is wasted, however, because cable TV systems today don't transmit bits; instead, they use analog technology to transmit thirty to seventy-five channels of video. Cable TV companies will continue to carry conventional television channels. But with the addition of new switches that support digital-information transmission, their cables will also be able to carry up to hundreds of millions of bits per second of information.

----------------------------➤

From a practical standpoint, the bandwidth a cable company provides to a particular home won't be as great as it might seem at first. Unlike telephone company wires, which can deliver a specific signal to a specific home, cable signals are broadcast to 200 to 1,000 homes. When the telephone company uses ADSL to feed a specific home 1.5 million bits per second, that is for the home's exclusive use. When the local cable company devotes 6MHz of bandwidth to the delivery of 27 million bits per second on a shared basis to a large neighborhood, all the cable modems compete for a share of the bandwidth. Either way, the home gets what is effectively a midband connection to the Internet. Phone companies could increase effective bandwidth by activating additional phone lines into homes, and cable companies could increase bandwidth by using more than one 6MHz channel for Internet connectivity—or by installing another fiber cable into the neighborhood.

Besides finding ways to offer Internet and conventional telephone service, another interim step cable companies will take is to increase the number of broadcast channels they carry five- to tenfold. They'll get the bandwidth they need for this increase by expanding their use of fiber and by using digital-compression technology to eliminate redundant information in the data stream, thereby squeezing more channels into whatever bandwidth is available.

Having 300 or 500 new channels makes near-video-on-demand possible, although for only a limited number of television shows and movies. You would choose from an on-screen list rather than select a numbered channel. A popular movie might be broadcast on twenty of the channels, with the starting times staggered so that you could begin watching the movie on one channel or another within five minutes of whenever you wanted to. You'd choose from among the available starting times for movies and TV programs, and the set-top box would switch to the appropriate channel. The half-hour-long *CNN Headline News*—or maybe a news program from MSNBC!—might be offered on six channels instead of one, with the 6:00 P.M. broadcast shown again at 6:05, 6:10, 6:15, 6:20, and 6:25. There would be a new, live broadcast every half hour, just as there is now. If some of the bandwidth is already funneled into Internet activity, the 500 channels' worth of bandwidth would get used up pretty fast this way.

The cable companies are under pressure to add channels partly as a reaction to competition. Direct-broadcast satellites such as Hughes Electronics' DIRECTV are already beaming hundreds of channels directly into homes (and are starting to broadcast data to PCs too). Cable companies want to increase their channel lineup rapidly to avoid losing customers.

If the only reason to have an interactive network were to deliver a limited number of movies, with or without some Internet connectivity, then a 500-channel system would be adequate. But a 500-channel system will still be mostly synchronous, will offer limited choices, and will provide only a low-speed back channel at best. A back channel on a 500-channel system might let you use your television set-top box to order products or programs, respond to polls or game-show questions, and participate in certain kinds of multiplayer games. But a low-speed back channel can't offer the full flexibility and interactivity that many interesting applications will require.

Don't get me wrong: Having 500 channels will be more interesting than what we have today, and I look forward to it. If the cable company can deliver both a midband Internet connection and hundreds of video feeds, that's even better. But it won't be the broadband information highway. It won't let you watch *any* past or current TV show any time you want to. It won't let you fast-forward, pause, or rewind any of thousands of movies delivered via the network to your home screen. It won't deliver a world of interactive educational and entertainment content that has high-quality video. It won't let your parents enjoy a high-quality home videoconference with their friends, children, or grandchildren.

It's unclear how many years it will take for millions of people to have broadband connections to the Internet. Fiber-optic lines and other enabling technologies will gradually reach neighborhoods as telephone companies and cable television companies routinely upgrade their systems. But having some or even all of the infrastructure in place is not the same thing as having an interactive broadband network. Applications and content are needed too. Millions of houses, apartments, and condominiums have to be connected up before the market will be big enough to justify the investment required to develop a wide array of interactive broadband content and applications aimed at the home.

---------------------------->

Fortunately, the hardware and software pieces of the broadband network don't have to come together all at once. The public's embrace of the Internet gives us an evolutionary path. The evolution is being driven already by the public's growing appetite for bandwidth. People using the Internet at home recognize that bandwidth limitations are constraining. They sense how much better the experience could be if the network connection were faster. People who have relatively high bandwidth connections at their offices or schools are wishing they could have those kinds of connections at home. This demand is encouraging telephone and cable companies to invest in midband access to the Internet for their residential customers.

Broadband connections will follow. The technology, applications, and content will be much more an outgrowth of the PC and the Internet and much less an outgrowth of the TV than many people expected them to be just a few years ago. This evolutionary path makes sense, and it's going to happen.

6

THE CONTENT
REVOLUTION

F or more than 500 years, the bulk of human knowledge and information has been stored as paper documents. You've got one in your hands right now (unless you're reading this from a CD-ROM or a future on-line edition). Paper will be with us for the foreseeable future, but its importance as a medium for finding, preserving, and distributing information is already diminishing.

When you think of a "document," you probably visualize one or more pieces of paper with print on them, but that's a narrow definition. A document can be any body of information. A newspaper article is a document, but the broadest definition of the word includes a Web page, a TV show, a song, or an interactive video game. Because all kinds of information can be stored in digital form, documents containing all kinds of information will get easier and easier to find, store, and send across a network. Paper is more awkward to store and transmit, and its content is pretty much limited to text with drawings and images. A digitally stored document can be made up of photos, video, audio, programming instructions for interactivity, animation, or a combination of these elements and others.

We'll be able to do things with these rich electronic documents we could never do with pieces of paper. The future network's powerful database technology will allow documents to be indexed and retrieved by means of interactive exploration. It will be extremely cheap and easy to distribute them. These new digital documents will replace many printed paper documents because they'll be able to help us in new ways.

But let's not sell paper documents short. The paper-based book, magazine, or newspaper still has a lot of advantages over its digital counterpart. A newspaper offers a wide field of vision, good resolution, portability, ease of use. A book is small, lightweight, high-resolution, and inexpensive compared to the cost of a computer or some other information appliance you need to read a digital document. For at least a decade, it won't be as convenient to read a long, sequential document on a computer screen as on paper. (I'll admit that I've done a lot of the editing of this book with a pen on paper. I *like* reading text on paper.) The first digital documents that achieve widespread use will offer new functionality rather than simply duplicate the older medium. After all, a television set is larger, more expensive, and more cumbersome and offers lower resolution than a book or a magazine, but that hasn't limited TV's popularity. Television brought video entertainment into our homes, and it was so compelling that television sets found their place alongside our books and magazines.

Ultimately incremental improvements in computer and screen technology will give us a lightweight, universal electronic book, or "e-book," that will approximate today's paper book. Inside a case roughly the same size and weight as today's hardcover or paperback book, you'll have a display for high-resolution text, pictures, and video. You'll be able to "flip" pages with your finger or use voice commands to search for the passages you want. Any document on the network will be accessible from such a device.

But the real point of electronic documents is not simply that we'll read them on hardware devices. Going from paper book to e-book is just the final stage of a process already well under way. The significant aspect of digital documents is the redefinition of the document itself, which will cause dramatic repercussions. We'll have to rethink not only what we mean by the term "document" but also what we mean by "author," "publisher," "office," "classroom," and "textbook."

It used to be that if two companies were negotiating a contract, the first draft was probably typed into a computer and then printed out on paper. Then it was faxed to the other party, who edited it by writing on the paper or reentering the document with some changes into another computer, from which the new version was printed out. Then the second party faxed the paper edit or a printout of the new version back to the first party; more changes were incorporated; a new paper document was printed out and faxed back again; and the editing process was repeated. During this transaction it was hard to tell who made which changes, and coordinating all the alterations and transmittals introduced extra overhead into the negotiations. With an electronic document, the process is simplified. The contract document goes back and forth in an instant, with corrections, annotations, and indications of who made them and when showing up along with the original text. Furthermore, the two parties are able to talk to each other over the network while they edit the document together.

Within a few years the digital document, complete with authenticatable digital signatures, will be the original, and paper printouts will be secondary. Many businesses have already advanced beyond paper and fax machines. They exchange editable documents, computer to computer, through electronic mail. This book would have been much harder to write without e-mail. Readers whose opinions I valued received drafts electronically, made electronic changes to the drafts, and sent the altered documents back to me. It was helpful to be able to look at the proposed revisions, see the rationales for the proposed changes in electronic annotations, and see the electronic record of who made the revision suggestions and when.

By the end of the decade a significant percentage of documents, even in offices, won't be fully printable on paper. The document will be like a movie or a song is today. You'll still be able to print out a two-dimensional view of its content, but that will be like reading a music score instead of listening to a recording.

Some kinds of documents are so superior in digital form that people don't feel the need for a paper version. Boeing designed its new 777 jetliner using a gigantic electronic document to hold all of the engineering information. During the development of each earlier airplane model,

Boeing had used blueprints and constructed an expensive full-scale mock-up of the airplane to coordinate collaboration among the design teams, manufacturing groups, and outside contractors. The mock-up had been necessary to make sure that the parts of the airplane, designed by different engineers, actually fit together the way they were supposed to. During development of the 777, Boeing did away with blueprints and the mock-up and from the start used an electronic document that contained digital 3-D models of all the parts and how they fit together. Engineers at computer terminals could look at the design and see different views of the parts. They could track the progress in any area, search for significant test results, annotate the design with cost information, and change any aspect of the design in ways that would be impossible on paper. Every person associated with the project, working with the same data, was able to look for what specifically concerned him. Every change could be shared, and everybody could see who made any change, when it was made, and why. Boeing was able to save hundreds of thousands of pieces of paper and many person-years of drafting and copying by using this digital document.

As you'd expect, digital documents can also be faster to work with than paper documents. You can transmit information instantly and retrieve information almost as quickly. People using digital documents now are already discovering how much simpler and quicker it is to search and navigate through them and how easy it is to restructure their content.

A reservation book at a restaurant is structured inflexibly by date and time. A 9:00 P.M. reservation is written farther down the page than an 8:00 P.M. reservation. Saturday night dinner reservations follow the reservations for Saturday lunch. A maître d' can quickly find out who has a reservation on any date for any time because the book's information is ordered by date and time. But if, for whatever reason, somebody wants to extract some other information—say, whether a particular person has a reservation at all, for any date and time—the simple chronology system is practically useless.

Imagine the plight of the restaurant captain if I called to say, "My name is Gates. I think my wife made us a reservation for some time next month. Would you mind checking to see whether she did and when it is?"

"I'm sorry, sir, do you know the date of the reservation?" the captain would probably ask.

"No, that's what I'm trying to find out."

"Would that have been on a weekend?" the captain asks.

He knows he's going to be looking through the book page by page, line by line, and he's hoping to narrow his search.

A restaurant can use a paper-based reservation book because the total number of reservations isn't that large and the captain wouldn't receive queries like mine that often. But an airline reservation system contains an enormous quantity of information—flight numbers, air fares, bookings, seat assignments, and billing information—for hundreds or thousands of flights a day worldwide. American Airlines' SABRE reservation system stores its information—4.4 trillion bytes of it, which is more than 4 million million characters—in a database on computer hard disks. If the information in the SABRE system were copied into a hypothetical paper reservation book, it would take more than 2 billion pages. Fortunately, American Airlines personnel don't have to leaf through 2 billion pages of reservation data. Anybody with access to the system can find any piece of information in it in several different ways.

For as long as we've had paper documents or collections of documents, we've been organizing information linearly, with indexes, tables of contents, and cross-references of various kinds to provide alternative means of navigation. In most offices, filing cabinets are organized by customer, vendor, or project in alphabetical order, but to speed access, a duplicate set of records is often filed chronologically. Professional indexers add value to a book by building in alternative ways to find information. And before library catalogs were computerized, new books were entered into the paper catalogs on several different cards so that a reader could find a book by its title, its author(s), or its topics. The redundancy made information easier to find.

When I was young, I loved my family's 1960 *World Book Encyclopedia*. Its heavy bound volumes contained just text and pictures. They showed me what Edison's phonograph looked like but couldn't let me listen to its scratchy sound. The encyclopedia had photographs of a fuzzy caterpillar changing into a butterfly, but there was no video to bring that transformation to life. It would have been nice too if the encyclopedia could have quizzed me on what I'd read, or if the information had always been up-to-date. I wasn't aware of those deficiencies then, though. When I was eight, I began to read the first volume, determined to read straight through every one. I could have absorbed more if it had been easy to read all the articles about the sixteenth century in sequence or all the articles pertaining to medicine in sequence. Instead I read about "Garter Snakes," then "Gary, Indiana," then "Gas." I had a great time reading the encyclopedia anyway and kept at it for five years, until I reached the Ps. Then I discovered the *Encyclopaedia Britannica*, with its greater sophistication and detail. I knew I'd never have the patience to read all of it, and by then, satisfying my enthusiasm for computers was taking up most of my spare time.

A current print encyclopedia consists of nearly two dozen volumes, with millions of words of text and thousands of illustrations, and it will cost hundreds or thousands of dollars. That's quite an investment, especially when you consider how rapidly the information goes out-of-date. One multimedia encyclopedia that comes on a single 1-ounce CD-ROM (Compact Disc—Read Only Memory) includes 26,000 topics, 9 million words of text, 8 hours of audio, more than 8,000 photographs and illustrations, more than 950 maps, 250 interactive charts and tables, and 100 animations and video clips. It contains 300,000 links that relate articles—and it costs less than $60 in the U.S. If you want to hear what the Egyptian "ud" (a musical instrument) sounds like, hear the 1936 abdication speech of Great Britain's King Edward VIII, or see an animation explaining how a machine works, the information is all there—and no paper-based encyclopedia will ever have that kind of information.

An article in a print encyclopedia is often followed by a list of articles on related subjects. To read one, you have to find the referenced article, which may be in another volume. With a CD-ROM encyclopedia, all you have to do is click on the reference to the article (a link) and the article will appear. In the on-line world, encyclopedias also feature articles with links to update articles and articles on related subjects—not just to other articles in the encyclopedia, but to articles and other kinds of information in sources all over the world. As the Internet and on-line reference works evolve, there will be no practical limit to the amount of detail you'll be able to explore on a subject that interests you. An encyclopedia on the net will be more than just a reference work. It will be like the library card catalog—a point of departure for exploration.

In the past information has been hard to locate. It's been almost impossible to find all the best information—including books, news articles, and film clips—on a specific topic. And it's been extremely time-consuming to assemble the information you can find. If you wanted to read biographies of all the recent Nobel Prize laureates, for example, compiling them could take an entire day. Electronic documents, however, are interactive. Request a kind of information, and the document responds. Refine your request or indicate that you've changed your mind, and the document responds again. Once you get used to this kind of system, you'll find that being able to look at information in different ways makes that information more valuable to you. The flexibility

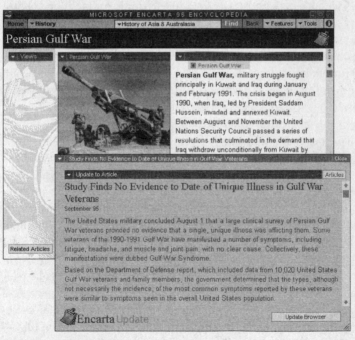

1996: Screen from on-line electronic multimedia encyclopedia, showing update article

invites exploration, and the exploration is invariably rewarded with discovery.

You'll be able to get your daily news in a similar way. You'll be able to specify how long you want your newscast to last. You'll be able to have each news story selected individually for you. The newscast assembled for just you might include world news from NBC, the BBC, CNN, or the *Los Angeles Times*, with a weather report from your favorite local TV meteorologist—or from any private meteorologist who offers his or her own service. You'll be able to ask for longer stories on the subjects that particularly interest you and for just highlights on others. If during the newscast you want more information than has been put together for you, you'll be able to ask for more background or detail, from either contemporary reports or file information.

Among all the types of paper documents, narrative fiction is one of the few that's not clearly improved by electronic organization. Almost every reference book has an index, but a novel doesn't. It is linear. Likewise, we'll continue to watch most movies from start to finish. This isn't a technological judgment—it's an aesthetic one. Linearity is intrinsic to the conventional storytelling process. New forms of interactive fiction that take advantage of electronic capabilities have been the subject of experimentation and may evolve into creative successes, but linear novels and movies will always be popular.

The network will make it easy to distribute digital documents cheaply, whatever their structure. Millions of people and companies will be creating documents and publishing them on the Web. Some documents will be aimed at paying audiences, and some will be free to anybody who wants to pay attention. Digital storage has become fantastically inexpensive. Hard disk drives in personal computers cost about $0.15 or less per megabyte (million bytes) of information storage. To put this in perspective, consider that 1 megabyte will hold about 700 pages of text, so the cost is something like $0.00021 per page—about one two-hundredth what the local copy center would charge at $0.05 a page. And because with digital storage you can reuse the storage space for something else later, the cost is actually the cost of storage per unit of time—in other words, the cost of renting the space. If we assume just a three-year average lifetime for the hard disk drive, the amortized price per page per year is $0.00007. And storage is getting cheaper all the time. Hard disk prices have been dropping by about 50 percent per year for the last several years.

Text is particularly easy to store because it's so compact in digital form. The decades-old saying that a picture is worth a thousand words is more than true but turned on its head in the digital world. High-quality photographic images take much more space than text, and video (which you can think of as a sequence of up to thirty new images appearing every second) takes even more. Nevertheless, the cost of distribution for even these kinds of data is still quite low. A feature film takes up about 4 gigabytes (4,000 megabytes) in compressed digital format, which is about $1,600 worth of hard disk space.

Sixteen hundred dollars to store a single film might not sound low-cost to you, but consider that the local video rental store might buy twenty copies of a hot new movie for $80 a copy. And with those twenty copies the store can supply only twenty customers per day.

Once the disk and the computer that manages it are connected up to the network, one copy of the information will be enough to give large numbers of people access. With one investment—roughly what a single shop spends today for its copies of a popular videotape title—a disk-based server will be able to serve thousands of customers simultaneously (although copies of really popular documents will be delivered from multiple servers to avoid delays when lots of people want to see them). The extra cost for each user is simply the cost of using the disk storage for a short period of time and the communications charge. And that's becoming extremely cheap. The extra per-user cost will soon be nearly zero.

This doesn't mean that information will be free, but the cost of distributing it will be very low. When you buy a paper book, a good part of your money goes for the cost of producing and distributing the book rather than for the author's work. Trees have to be cut down, ground into pulp, and turned into paper. The book has to be printed and bound. Most publishers invest capital in a first printing of the largest number of copies they think will sell right away because the printing technology is efficient (cost effective) only if lots of books are made in one print run. The capital tied up in this inventory is a financial risk for the publisher: They may never sell all the copies, and even if they do, it will take a while. Meanwhile, the publisher has to store the books and ship them to wholesalers and ultimately to retail bookstores. Those folks also invest capital in their inventory and expect a financial return from it.

By the time the consumer decides to buy the book and the cash register rings, the profit for the author can be a pretty small piece of the pie compared to the money that goes to the people who manage the physical aspects of delivering information on processed wood pulp—to the publisher, the printer, the distributor, and the bookseller. This is the "friction" of distribution, a drag on variety and a diversion of money away from the author and to other people.

Publishing via a network is largely friction-free, a theme I'll explore further in chapter 8. This lack of friction in information distribution is an important development. It will empower more authors because very little of the customer's dollar will pay for printing and distribution.

Gutenberg's invention of the printing press brought about the first real easing of distribution friction—it allowed information on any subject to be distributed quickly and relatively cheaply. The printing press created a mass medium because it offered low-friction duplication. The proliferation of books motivated the general public to read and write, and once people had those skills, they could do many other things with the written word. Businesses could keep track of inventory and write contracts. Lovers could exchange letters. Individuals could keep notes and diaries. By itself, none of these applications would have been sufficiently compelling to get large numbers of people to make the effort to learn to read and write. Until there was a real reason for an "installed base" of literate people to arise, the written word wasn't really useful as a means of storing information. Books gave literacy critical mass, so you can almost say that the printing press taught us to read.

The printing press made it easy to make lots of copies of a document, but what about something written for more than one but only a few people? For small-scale publishing, a new technology was required. Carbon paper was fine if you wanted just one or two more copies. Mimeographs and other messy machines could make dozens of copies, but to use any of these processes you had to plan for them when you prepared your original document.

In the 1930s Chester Carlson was frustrated by how difficult it was to prepare patent applications, which involved copying drawings and text by hand. He set out to invent a better way to duplicate information in relatively small quantities and came up with the process he called "xerography" when he got a patent for it in 1940. In 1959 the company he had hooked up with—later known as Xerox—released its first successful production line copier. The 914 copier, by making it possible to reproduce modest numbers of documents easily and inexpensively, set off an explosion in the kinds and amounts of information distributed to small groups. In their market research, Xerox had projected that they'd sell at most 3,000 of their first copier model. They actually placed about 200,000, and a year after the copier was introduced, businesses were making 50 million copies a month. By 1986 more than 200 billion copies were being made each month, and the number has been rising ever since. Most of these copies would never be made if the technology weren't so cheap and easy.

The photocopier and its younger cousin, the desktop laser printer—along with PC desktop publishing software—facilitated the production of newsletters, memos, maps to parties, flyers, and other documents intended for modest-size audiences. Carlson had reduced information distribution friction.

Despite the friction of distribution, there's no intrinsic limit to the number of books that can be published in any year, so lots of interesting titles get published—and get a shot at success. A typical bookstore has 10,000 different titles at any given time, and some superstores might carry 100,000. Only a small fraction, less than 10 percent, of all trade books published makes money for its publishers, but some books succeed beyond anybody's wildest expectations.

My favorite example is *A Brief History of Time*, by Stephen W. Hawking, a brilliant scientist who has amyotrophic lateral sclerosis (Lou Gehrig's disease), which confines him to a wheelchair and allows him to communicate only with great difficulty. What are the odds that Hawking's treatise on the origins of the universe would have been published if there had been only a handful of publishers and each of them could have produced only a few books a year? Suppose an editor had had one spot left on his list and had had to choose between publishing Hawking's book and Madonna's *Sex*? The obvious bet would have been Madonna's book because it would have looked likely to sell a million copies. It did. But Hawking's book sold 5.5 million copies and is still selling.

Every now and then a sleeper bestseller surprises everybody (except the author). Nobody really knows what will appeal to the public's taste. There are almost always a couple of books on the *New York Times* bestseller list that have bubbled up from nowhere. Books cost so relatively little to publish—compared to most other media—that publishers can afford to give lots of them a chance.

Costs are much higher in broadcast TV or the movies, and many fewer titles can be distributed, so it's tougher to try something risky. In the early days of TV there were only a few stations in each geographic area, and most programming was targeted for the broadest possible audience.

Cable television increased the number of programming choices, although the people who started it up didn't have programming diversity in mind. Cable TV was developed in the late 1940s as a way of providing better TV reception to outlying areas. Viewers whose broadcast reception was blocked by hills put up community antennas to feed a local cable system. Nobody imagined then that people with perfectly good broadcast TV reception would pay to have cable run into their homes so that they could watch a steady stream of music videos or a channel that offered nothing but news or nothing but weather twenty-four hours a day.

When the number of stations a viewer could choose from went from three or five to twenty-four or thirty-six, the TV programming dynamic changed. If you were in charge of programming for the thirtieth channel, you wouldn't attract much of an audience if you just tried to imitate channels 1 through 29. Instead, cable channel programmers were forced to specialize. Like special-interest magazines and newsletters, these new channels attract viewers by appealing to strong interests held by relatively small numbers of enthusiasts—in contrast to general programming, which tries to provide something for everybody. This logic of specialization can take us only so far, though. The costs of production and the small number of channels still limit the number of television programs that get produced.

Although it costs far less to publish a book than to broadcast a TV show, publishing a book still costs a lot compared to publishing on the Internet. To get a book into print, an author generally has to find a publisher who will agree to pay the up-front costs of manufacturing, distribution, and marketing. But the Internet is a publishing medium with entry barriers lower than any we've ever seen—the greatest self-publishing vehicle ever. Its proliferation of bulletin boards, newsgroups, and Web pages demonstrates the changes that take place when millions of people have access to low-friction distribution and anybody can post messages, images, or software.

Bulletin boards and newsgroups contributed a lot to the popularity of the early Internet. To be published, all you had to do was type in your thoughts and post them. This did and does lead to a lot of garbage on the Internet, but also a few gems. A typical message is only a page or two long. A single message posted on a popular bulletin board or sent to a mailing list might reach and get the attention of millions of people. Or it might sit there, having no impact whatsoever. The reason any author/publisher is willing to risk the latter eventuality is the low distribution friction. The network bandwidth is so great and the costs are so low that nobody thinks about the cost of sending a message. At worst you might be a little embarrassed if your message just sits there and nobody responds to it. At best a lot of people will see your message, forward it as e-mail to their friends, and post their own comments in response to it.

It's fast and inexpensive to communicate over the Internet. Mail or telephone communications are fine for a one-on-one discussion, but they get pretty expensive if you're trying to communicate with a group. It costs nearly a dollar to print and mail just one letter and on average about that much for a long-distance call. And to make a conference call you have to spend even more, assemble all the phone numbers, and have coordinated a time when everybody will be free to talk with the other people. It takes considerable time and effort to put even a modest-size group of people in touch with each other. On a bulletin board, all you have to do is type your message in once and it's available to everybody you want to reach.

Bulletin boards on the Internet cover a wide range of topics, and postings don't have to be serious. Somebody will send a humorous message to a mailing list or post it somewhere. If it's funny enough, it will get forwarded to innumerable people and groups as e-mail. In late 1994 this happened with a phony press release about Microsoft's acquiring the Catholic Church. Thousands of copies were distributed inside Microsoft on our e-mail system. I received more than twenty copies myself as various friends and colleagues inside and outside the company forwarded them to me.

Bulletin boards and e-mail have been used to mobilize people who share a serious common concern or interest too. During political conflicts in Russia, both sides were able to contact people throughout the world via postings on electronic bulletin boards. The Internet lets you contact people you have never met who share an interest with you.

Information published by electronic posting is grouped by topic. Each bulletin board or newsgroup has a name, and anybody who's interested can "hang out" there. You can find lists of interesting newsgroups, or you can browse bulletin boards with names that sound interesting to you. If you wanted to get in touch with a group interested in paranormal phenomena, you'd go to the newsgroup alt.paranormal. If you wanted to discuss that sort of thing with people who don't believe in it, you'd go to sci.skeptic. You could connect to copernicus.bbn.com and look in National School Network Testbed for a set of lesson plans used by kindergarten through twelfth-grade teachers. Almost any topic you can name has a group communicating about it on the network.

Gutenberg's invention started mass publishing, but the rise in literacy it stimulated ultimately led to a great deal more person-to-person correspondence. Computer-based communication across networks developed the other way around. It started out as electronic mail, a way to communicate person-to-person or to small groups, and developed into a form of mass communication. Now millions of people are taking advantage of the Internet's low-friction distribution to communicate on a wide scale via various forms of posting.

Perhaps the most powerful form of posting, the one that most resembles a traditional publication, is the Web page. The explosion of content on the Internet has taken place on its World Wide Web. To make it easy for nontechnical people to create their own Web home pages, freeware or shareware software "wizard" applications are available to guide people through a form that asks for the user's name and hobbies, favorite sites on the Internet, a favorite saying, and images of the user's children or maybe a favorite car or the family dog. Then the wizard creates a personalized page that the user can upload to his own Internet provider such as America Online or UUNet. Presto! Instant worldwide publishing.

The quality of Web publishing is uneven, as you'd expect in a medium in which anybody can publish and nobody can easily charge anything for their work. I hate to think how many hours I've spent cruising around, looking at information of dubious value. But as millions of people have discovered, browsing the Web can be enthralling even so. There's so much content, from so many sources, that companies have sprung up to publish Web pages that are nothing more than roadmaps to other Web pages—and some of these fledgling directory companies have gone public with valuations in the hundreds of millions of dollars. I described one of these services, YAHOO, in Chapter 4.

The interactive network has enormous potential, but it's important for its continuing credibility that expectations don't get cranked too high. The total number of users of the Internet and of commercial online services such as CompuServe, America Online, and the Microsoft Network (MSN) is still a small part of the population. And the attrition rate for the on-line services is high—many subscribers drop off, disappointed, after less than a year. Others shop around and change services every couple of months or so to take advantage of special offers. Some services, such as Apple's e-World, have shut down. Others, such as Prodigy, have experienced discouraging results and face an uncertain future. Microsoft's MSN attracted more than a million subscribers in its first seven months of operation, but the skyrocketing popularity of the Internet forced us to recast MSN as a service that embraces and extends the Internet rather than as an alternative to it. Other on-line services are doing the same thing.

Significant investments will be required to develop great on-line content that will attract PC users and raise the number on-line from 10 percent to 50 percent to the 90 percent I foresee. Content development has been held back somewhat because the simple, secure mechanisms authors and publishers need for charging users or advertisers are just beginning to appear.

Commercial on-line services collect revenue, but they have been paying information providers royalties of only 10 percent to 30 percent of what customers pay them. Although the content provider probably knows the customers and market better, pricing and marketing have both been controlled by the service. The resulting revenue stream simply hasn't been big enough to encourage a large number of information providers to create exciting new on-line information and keep it up-to-date.

Over the next several years the evolution of on-line service mechanisms will solve these problems and create an incentive for providers to furnish exciting, made-for-the-medium content. New billing options—monthly subscriptions, hourly rates, charges per item accessed, and advertising payments—will get more revenue flowing to the information providers.

Whenever a new medium comes on the scene, its early content is brought over from other media. But to take best advantage of the capabilities of the electronic medium, content needs to be specially authored with the medium in mind. Up until just recently most of the on-line content we've seen has been "dumped" from another source. Magazine and newspaper publishers took text already created for paper editions and simply shoved it on-line, often minus the pictures, charts, and graphics. This content was often interesting but it couldn't compete with the richer forms of information in our lives. Now most on-line content from commercial print publishers includes lots of graphics, photos, and links to related information. As communications get faster and commercial opportunities become even more apparent, more audio and video elements will be brought into on-line content.

CD-ROMs provide us with some models for the creation of on-line content. CD-ROM-based multimedia titles have integrated different types of information—text, graphics, photographic images, animation, music, and video—into single documents, and right now they're our best approximations of what the rich documents of the future will be like.

The music and other audio on CD-ROMs is clear but rarely as good as on a music CD. You could store CD-quality sound on a CD-ROM, but the format a CD-ROM uses for audio is bulky. If you stored much CD-quality sound, you wouldn't have room for data, graphics, and other kinds of media.

Motion video on CD-ROMs needs improving too. Longtime computer users got excited when they first encountered video on their computers, but the grainy, jerky image was certainly no better than a 1950s television picture. If you compare the quality of video a PC from 1996 can display with the postage-stamp-size video images earlier in the decade, the progress is remarkable. But we're still not where we should be. The size and quality of images will improve with faster processors and better compression, and eventually images on the PC will be far better than today's television picture.

Even with its limitations, though, CD-ROM technology has brought about new categories of applications that will influence on-line content. Shopping catalogs, museum tour guides, and textbooks are being published in this appealing multimedia form. Subjects of every kind are getting multimedia treatment, and competition and advances in technology will bring rapid improvements in the quality of the titles. The CD-ROM will be replaced over time by the digital video disk (DVD), a new high-capacity disc that resembles a conventional CD but holds ten times as much data. The additional capacity of these extended CDs allows for more than two hours of digital video, enough for a movie. The picture and sound quality will be much higher than the quality of the best TV signal you can receive on a home set, and new generations of graphics chips will allow multimedia titles to include Hollywood-quality special effects under the interactive control of the user.

Multimedia CD-ROMs are popular today not because they have imitated TV but because they offer users interactivity. The commercial appeal of interactivity has already been demonstrated by the popularity of CD-ROM games such as Brøderbund's *Myst* and Virgin Interactive Entertainment's *Seventh Guest*, whodunits that blend narrative fiction with a series of puzzles that enable a player to investigate a mystery, collecting clues in any order.

The success of these games has encouraged authors to create interactive novels and movies in which they introduce the characters and the general outline of the plot and the reader/player then makes decisions that change the progress and the outcome of the story. No one would suggest that every book or movie should allow the reader or viewer to influence its outcome. A good story can make you just want to sit back for a few hours and enjoy. I don't want to choose an ending for *The Great Gatsby* or *La Dolce Vita*. F. Scott Fitzgerald and Federico Fellini have done that for me. The suspension of disbelief essential to the enjoyment of great fiction is fragile enough that it might not hold up under a heavy-handed use of interactivity. You can't simultaneously control the plot and surrender your imagination to it.

Interactive stories and games will be available on the network too. CD-ROMs can share their content with network applications, but for at least a while the software will have to be designed so that the CD-ROM content won't be slowed down when it's used on the network. The speed at which bits are transferred from the CD-ROM to the computer—the bandwidth—is far greater than the bandwidth of our existing telephone network. As high-speed connections to the Internet become common, the content created for the two media can be the same. But this compatibility will take a number of years to achieve, especially considering that improvements are also being made in CD-ROM-DVD technology.

Other than Web pages, very few multimedia documents are being created by PC users so far. It still takes too much effort. Millions of people have camcorders and make videos of their kids or their vacations. But to edit video right now you have to be a professional with expensive equipment. This will change. Advances in PC word processors and desktop publishing software have already made professional-quality tools for creating sophisticated paper documents available relatively inexpensively to millions of people. Desktop publishing software has progressed to the point that many magazines and newspapers are produced with the same sort of PC and software packages you can buy at any local computer store and use to design an invitation to your daughter's birthday party. PC software for editing film and creating special effects will become as commonplace as desktop publishing software. At that point the difference between professionals and amateurs will be one of talent and craft rather than access to tools.

Georges Méliès created one of the first special effects in movies in 1899, when he turned a woman into feathers in *The Conjurer*, and moviemakers have been playing cinematic tricks ever since. Recently special effects technology has improved dramatically with the digital manipulation of images. First a photograph is converted into binary information, which, as we've seen, software applications can easily manipulate. Then the digital information is altered and finally returned to photographic form, as a frame in a movie. The alterations are nearly undetectable if they're well done, and the results can be spectacular. Computer software gave life to the dinosaurs in *Jurassic Park*, the rival toys Woody and Buzz in *Toy Story*, and the crazy cartoon effects in *The Mask*. Many of the hit movies of 1996, from *Twister* to *Independence Day*, relied heavily on special effects. As hardware speed and software sophistication increase, there's virtually no limit to the special effects that can be achieved. Hollywood will continue to push the state of the art.

It's already possible for a software program to fabricate scenes that look as real as anything recorded with a camera. Audiences watching *Forrest Gump* knew that the scenes in which Gump appears with Presidents Kennedy, Johnson, and Nixon were fabricated. Everybody knew that actor Tom Hanks hadn't really been there. It was a lot harder to spot the digital processing that removed Gary Sinise's two good legs for his role as an amputee. Synthesized figures and digital editing are also being used to make movie stunts safer. You'll soon be able to use a standard PC to create the effects. The ease with which PCs and photo-editing software already manipulate complex images makes it easy to counterfeit photographic documents or alter photographs undetectably. As synthesis gets cheaper, it will be used more and more. If we can bring Tyrannosaurus rex back to life, can Elvis be far behind?

Even people who don't aspire to becoming the next C. B. DeMille or Lina Wertmuller will routinely construct multimedia documents every day. Someone might start by typing, handwriting, or speaking an electronic mail message: "Lunch in the park may not be such a great idea. Look at the forecast." To make the message more informative, he might point his cursor at an icon representing a local television weather forecast and drag it across his screen to drop it inside his document. When his friends get the message, they'll be able to look at the forecast right on their screens—a professional-looking communication.

Kids in school will be able to produce their own albums or movies and use the net to make them available to friends and family. When I have time, I like to make special greeting cards and invitations. If I'm making a birthday card for my sister, for instance, to personalize it I sometimes add pictures reminding her of things we did that were fun in the last year. For Christmas 1995, I gave many of my relatives Kodak digital cameras that record images electronically instead of on film. It's easy to use a PC to edit out "red eye" and other imperfections, and now we use the Internet to exchange our photos across the many miles that separate some of us. In the future we'll be able to customize movie clips with only a few minutes' work. It will be simple to create an interactive "album" of photographs, videos, or conversations. Businesses of all types and sizes will communicate using multimedia. Lovers will use special effects to blend some text, a video clip from an old movie, and a favorite song to create a personal valentine. Ex-lovers will use digital tools to delete each other from photographs. (Don't laugh—it's happening already.)

As the fidelity of visual and audio elements improves, we'll be able to more closely simulate many aspects of reality. This "virtual reality," or VR, will enable us to "go" places and "do" things we never would be able to otherwise.

Vehicle simulators for airplanes, race cars, and spacecraft already give us a taste of virtual reality. Some of the most popular rides at Disneyland are simulated voyages. Software simulators such as flight simulators are among the most popular games ever created for PCs, but they force you to use your imagination. A multimillion-dollar flight simulator at Boeing can give you a much more realistic ride. Seen from the outside, it's a boxy, stilt-legged mechanical creature that would look at home in a *Star Wars* movie. Inside, the cockpit video displays give you sophisticated data. Flight and maintenance instruments are linked to a computer that simulates flight characteristics—including emergencies—with an accuracy that pilots say is remarkable.

A couple of friends and I "flew" a 747 simulator a few years ago. You sit down to a control panel in a cockpit identical to one in a real plane. Outside the windows, you see computer-generated color video images. When you "take off" in the simulator, you see an identifiable airport and its surroundings. The simulation of Boeing Field, for instance, might show a fuel truck on the runway and Mount Rainier in the distance. You hear the rush of air around wings that aren't there, the clunk of nonexistent landing gear retracting. Six hydraulic systems under the simulator tilt and shake the cockpit. It's pretty convincing.

The main purpose of these simulators is to give pilots a chance to get experience in handling emergencies. When I was using the simulator, my friends decided to give me a surprise by having a small plane fly by. While I sat in the pilot's seat, the all-too-real-looking image of a Cessna flashed into view. I wasn't prepared for the "emergency," and I crashed into the Cessna.

A number of companies, from entertainment giants to little startups, are putting smaller-scale simulator rides into shopping malls and urban sites. As the price of technology comes down, entertainment simulators may become as common as movie theaters are today. And it won't be too many years before you'll be able to have a high-quality simulation in your living room.

Want to explore the surface of Mars? It's a lot safer to do it via VR. How about visiting places humans will never be able to go? A cardiologist might be able to swim through the heart of a patient to examine it in a way she never could have with conventional instrumentation. A surgeon might practice a tricky operation many times, with simulated catastrophes, before she ever touches a scalpel to a real patient. Or you could use VR to wander through a fantasy of your own design.

VR relies on software to create a scene and make it respond to new information and on hardware that allows the computer to transmit the information to our senses. The software will have to describe the look, sound, and feel of the artificial world down to the smallest detail. That might sound overwhelmingly difficult, but actually it's the easy part. Although we need a lot more computer power to make the results truly believable, at the pace technology is moving the power will be available soon. The really hard part about VR is outputting the information in a way that convinces the user's senses.

-----------------------------►

Hearing is the easiest sense to fool. All you have to do is wear headphones. In real life, your two ears hear slightly different things because of their location on your head and the directions to which they're attuned. Subconsciously you use those differences to tell where a sound is coming from. Software can re-create the differences by calculating for a given sound what each ear would be hearing. This works amazingly well. You can put on a set of headphones connected to a computer and hear a whisper in your left ear or footsteps walking up behind you.

Your eyes are harder to fool. VR equipment almost always includes a special set of goggles with lenses that focus each eye on its own small computer display. A head-tracking sensor enables the computer to figure out which direction you're facing so that the computer can synthesize what you'd be seeing. Turn your head to the right, and the scene the goggles show you is farther to the right. Lift your face, and the goggles show the ceiling or the sky. Today's VR goggles are too heavy and too expensive, and they don't have fine-enough resolution. The computer systems that drive them are still a little slow too. If you turn your head quickly, the scene lags behind somewhat. The lag is disorienting and after a short period of time gives most people a headache. But the technology is improving rapidly.

Other senses are much more difficult to fool because there's no good way to connect a computer to your nose or your tongue. In the case of touch, current thinking is that a full bodysuit could be lined with tiny sensor and force feedback devices that would be in contact with the whole surface of your skin. I don't think bodysuits will be common, but they'll be feasible.

There are between 72 and 120 tiny points of color (called pixels) per inch on a typical computer monitor, for a total of between 300,000 and 1 million. A full bodysuit would presumably be lined with little touch sensor points—each of which could poke one specific tiny spot. Let's call these little touch points "tactels."

If the suit had enough of these tactels, and if they were controlled finely enough, any touch sensation could be duplicated. If a large number of tactels poked all together at precisely the same depth, the resulting "surface" would feel smooth, as if a piece of polished metal were against your skin. If they pushed with a variety of randomly distrib-uted depths, it would feel like a rough texture.

Between 1 million and 10 million tactels—depending on how many different levels of depth the tactels had to convey—would be needed for a VR bodysuit. Studies of the human skin show that a full body-suit would have to have about 100 tactels per inch—a few more on the fingertips, lips, and a couple of other sensitive spots. Most skin actually has poor touch resolution. I'd guess that 256 levels of pressure per tactel would be enough for the highest-quality simulation. That's the possible number of colors for each pixel on many computer displays.

The total amount of information a computer would have to calculate to pipe the sense of touch into the tactel suit is somewhere between one and ten times the amount required for the video display on a current PC. This really isn't a lot of computer power. I'm confident that as soon as someone makes the first tactel suit, PCs of that era will have no problem driving them.

Sound like science fiction? The best descriptions of VR actually come from cyberpunk science fiction like William Gibson's. Rather than put-ting on a bodysuit, some of Gibson's characters "jack in" by plugging a computer cable directly into their central nervous systems. It will take scientists a while to figure out how this can be done, and when they do, it will be long after the broadband interactive network has been estab-lished. Some people are horrified by the notion, whereas others are intrigued. Such an innovation would probably be used initially to help people with physical disabilities.

Inevitably, there has been more speculation (and wishful thinking) about virtual sex than about any other use for VR. Sexually explicit con-tent is as old as information itself, and it never seems to take long to figure out how to apply any new technology to the oldest desire. The Babylonians left erotic poems in cuneiform on clay tablets, and sexually explicit material was one of the first things the printing press was used for. When VCRs became common home appliances, they provoked a surge in the sales and rentals of X-rated videos. Today sexually explicit CD-ROMs are popular, and the Internet's World Wide Web has many sites devoted to sexually explicit topics. If historical patterns are a guide, then, a big early market for advanced virtual reality documents will be virtual sex. But historically again, as each of the earlier markets grew, sexually explicit content became a smaller and smaller factor.

Imagination will be a key element in creating content for all new applications. It isn't enough just to re-create the real world. Great movies are a lot more than just graphic depictions on film of real events. It took a decade or so for such innovators as D.W. Griffith and Sergei Eisenstein to take the Vitascope and the Lumières' Cinématographe technology and figure out that motion pictures could do more than record real life or even a play. Moving film was a new and dynamic art form, and the way it could engage an audience was very different from the way the theater could. The pioneers saw this and invented movies as we know them today.

Will the next decade bring us the Griffiths and Eisensteins of multimedia? There's every reason to think they're already tinkering with the existing technology to see what it can do and what they can do with it.

I expect multimedia experimentation to continue into the next decade, and the one after that, and so on indefinitely. The multimedia components appearing in documents on the net today are a synthesis of current media—and they often do a clever job of enriching communication. But over time we'll start to create new multimedia forms and formats that will enable us to go significantly beyond what we're able to do now. The exponential expansion of computing power will keep changing the tools and opening up new possibilities that might seem as remote and farfetched then as some of the things I've speculated about here might seem today. Talent and creativity have always shaped advances in unpredictable ways.

How many have the talent to become a Steven Spielberg, a Jane Austen, or an Albert Einstein? We know there has been at least one of each, and maybe one is all we're allotted. I can't help but believe, though, that the potential and the aspirations of many talented people have been thwarted by economics and a lack of tools. New technology will offer people new means with which to express themselves. The Internet will open up undreamed-of artistic and scientific opportunities to a new generation of geniuses—and to everybody else, too.

7

BUSINESS ON
THE INTERNET

O ver the next decade, businesses worldwide will be transformed.
Intranets will revolutionize the way companies share information
internally, and the Internet will revolutionize how they communicate
externally. Corporations will redesign their nervous systems to rely on
the networks that reach every member of the organization and beyond
into the world of suppliers, consultants, and customers. These changes
will let companies be more effective and often smaller. In the longer run,
as broadband networks make physical proximity to urban services less
essential, many businesses will decentralize and disperse their activities,
and cities may be downsized too.

Businesses welcome information technology because long-term success in business depends on improving productivity. Network connections and a greater use of electronic documents promise companies
benefits such as Web publishing, videoconferencing, e-mail, flexible
ways of viewing data, and easier collaboration among staff, suppliers,
and customers worldwide. Even the smallest businesses will share in the
commercial advantages of information technology.

Some companies are already using electronic communication over the Internet to stay closer to their customers, a trend that will accelerate with far-reaching results. Although broadband interactive service won't reach many homes for years, telephone and cable companies will rush to connect businesses to interactive broadband networks in urban areas, where the concentration of business customers could mean early profits. Software applications aimed at corporate users have begun to take advantage of these high-speed connections already.

But in the near future the development that will help businesses the most will be the widespread creation and use of intranets—private Internets used to share information inside organizations. Intranets build directly on the investments in information technology that companies have already made. Even if they don't have superfast connections to the outside world, most large and medium-size companies have internal networks that provide midband or even broadband connections between their PCs. Traditional corporate Ethernet networks run at 10 million or 100 million bits per second, but since the bandwidth is usually shared by a lot of computers, the effective bandwidth—the number of bits that can be delivered to a particular PC at a particular time—is considerably lower than this. Still, corporate data delivery is a whole lot faster than the service that people at home with telephone dial-up modems are accustomed to. In the next few years we'll see even bigger improvements in the amount of bandwidth delivered to corporate computers as companies upgrade to faster Ethernet and ATM equipment.

Beginning in the 1980s companies made huge expenditures for personal computers and the networks that connect them. People learned new ways of creating documents with word processors, database packages, spreadsheets, presentation graphics applications, and other software tools. Without question, these investments have extended the capabilities of individuals and have made for better information sharing inside companies.

Along the way, there has been quite a shift in the way we think about and use computers as business tools. When I was a kid, my image of computers was that they were very big and powerful. Banks had lots of them, and big airlines used them to keep track of reservations. They were the big tools of large organizations and contributed to the edge big businesses had over the small guys who used pencils and typewriters.

Then the personal computer appeared. As its name implies, it is a tool for the individual, even in a large company. You use a personal computer in a very personal way to get a job done. People doing solo work can write reports, memos, and correspondence, keep the books, bill customers, construct profit and loss statements, create newsletters, design products, control manufacturing, and explore new business ideas better with a personal computer.

Businesses of all sizes benefit from personal computers, but small companies are arguably the greatest beneficiaries. Low-cost hardware and software allow even a tiny outfit to compete effectively, in its area of focus, with large multinational corporations. Big organizations tend to have specialists: One department writes marketing brochures, another deals with accounting, yet another handles customer service, and so on. When you call a large company to talk about your account, you expect a specialist to get you an answer fast. Your expectations of a small business are somewhat lower because small outfits can't afford specialists. When an individual runs a business or a shop, she is the one who creates the brochures, keeps the accounts, and deals with customers. I sometimes marvel at how many different tasks a small business owner has to master and then do, day after day. A personal computer can help a great deal. Someone running a small operation with a PC and a few software packages has electronic support for all the different functions she needs to perform. She can track customers, analyze sales, and create marketing materials as effectively as a large company. Her Web pages, for example, can look as professional as a giant corporation's, although they may not be as extensive. Thanks to the personal computer, a small business can compete more effectively with the big players than at any time in the past.

Even the smallest of all businesses, the individual earning a living in a profession or as an artist, has been empowered by the PC. One person without any staff can produce reports, handle correspondence, bill customers, and maintain a credible business presence—all surprisingly easily. In field after field, the tools of the trade have been transformed by PCs and software. The architect often does his drafts and renderings on a

PC. The surveyor's productivity has increased because software now does much of the once time-consuming calculation necessary to determine property lines—and a printer connected to a PC spits out the boundary and topographical maps. The composer uses software to print the musical score, which a musician may perform on a computer-based synthesizer. The screenwriter worries less about formatting screenplays and more about the quality of the dialog. Writers of all kinds have been major beneficiaries of the PC.

A skeptic might ask, "If Churchill had used a word processor, would his writing have been better? Would Cicero have given better speeches in the Roman Senate?" Such critics point out that great things were achieved without modern tools and question whether better tools can elevate human potential. It's hard to speculate on what might have been where genius on the order of a Da Vinci's or a Shakespeare's is concerned, but it's quite clear that personal computers improve the efficiency and accuracy of even very talented people. There have been great journalists throughout history, but today it's much easier to check facts, transmit a story from the field while it's still fresh, and stay in touch electronically with news sources, editors, and even readers. Production values in journalism are rising too. I've been following science news since I was a kid, and until the PC came along the only places I'd ever found top-notch scientific illustration had been in science books and glossy magazines like *Scientific American*. Now, thanks to illustration software running on PCs, high-quality, detailed drawings are commonplace in daily newspapers.

The biggest benefit from personal computers in large companies is an improvement in the sharing of information. Connected PCs reduce the overhead large businesses incur just staying coordinated through meetings, reports, memos, policy statements, and procedure manuals. Electronic mail, for instance, does much more for the internal communications of big companies than it does for small companies.

Even before the emergence of the Internet some businesses in the United States were already exchanging information via an electronic system called Electronic Document Interchange. EDI allows companies that have contractual relationships to execute specific kinds of transactions automatically, often using proprietary networks. Dealings are highly structured, which makes EDI suitable for reordering products, checking the status of a shipment, or other prearranged forms of communication. Today's EDI is unsuitable for ad hoc communication, though. Many companies are working to combine the benefits of EDI with Internet-based e-mail, to create systems that are more flexible and less expensive. It's unclear whether EDI will evolve fast enough or whether other standards that are richer will emerge. It is clear that the Internet will be the network used for electronic business-to-business communications.

The trend toward the Internet and away from proprietary data-communications systems such as leased telecommunications lines extends to branch-office corporate communications too. Global businesses have been spending heavily on private wide-area networks, but soon the Internet will connect all of a company's offices instead. Microsoft's subsidiary in Greece has been paying more for its connection to our private network than it pays in salaries. The Internet will be both faster and less expensive for these kinds of connections.

Companies embrace information technology at different rates, and some businesses are far ahead of others. But before a manager invests, he or she should remember that a computer or a network of computers is just a tool to help solve identified problems. It isn't, as businesspeople sometimes seem to expect it to be, a cure-all. If I heard a business owner say, "I'm losing money, I'd better get some computers," I'd tell him to rethink his strategy before he invests. Technology would at best probably only delay the need for more fundamental changes. The first principle for any technology you contemplate introducing into a business is that automation applied to an efficient operation will magnify the efficiency. The second is that automation applied to an inefficient operation will just entrench the inefficiency.

Instead of rushing out to buy the latest and greatest equipment for every employee or investing in a network, managers in a company of any size should first step back and think about how they'd like their business to work. What are its essential processes and its key databases? Ideally, how should information move?

In your company, for example, when a customer calls, does all of the information about your dealings with the customer—the status of the account, any complaints, who in your organization has worked with the customer—immediately appear on a screen? The technology for summoning up this kind of information is quite straightforward, and customers increasingly expect the level of service it affords. Some car companies now centralize service information so that any dealer can easily check a vehicle's entire service history and watch for recurring problems. If your systems can't provide on-the-spot product availability information or an immediate price quote, you risk losing out to a competitor who's taking better advantage of technology.

A company should also examine all of its internal processes—employee reviews, business planning, sales analysis, product development—to determine how computers, networks, and other electronic information tools can make these operations more effective.

Conventionally, businesses share information internally by exchanging paperwork, making telephone calls, and gathering around conference tables and white boards. You need plenty of time and plenty of expensive face-to-face meetings and presentations to reach good decisions that way. The potential for inefficiency is enormous. Companies that continue to rely on these methods exclusively risk losing out to competitors who reach decisions faster while devoting fewer resources, and probably fewer layers of management, to the process.

At Microsoft, because we're in the technology business, we began using electronic communication early—back when we were a little company. We installed our first e-mail system in the early 1980s. Even when we had only a dozen employees, e-mail made a difference, and it quickly became our principal method of internal communication. We used e-mail in place of paper memos and many meetings, to set up the meetings we still wanted to have, for quick technology discussions, for trip reports, for phone messages, and for reaching consensus of all kinds—from when we'd be able to deliver our next product to what kinds of toppings we wanted on a pizza. E-mail contributed a lot to our efficiency when we were small, and it's essential to us now that we have thousands of employees. Without it, we couldn't move as fast as we do.

E-mail is easy to use. To write and send an electronic message, I click on a button that brings a simple form up on the screen. First I type in the name of the person or people I want to address the message to, or I choose the

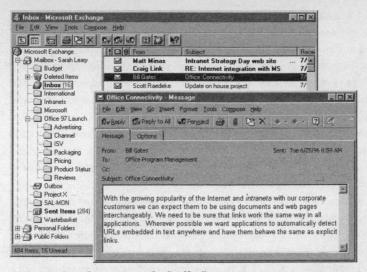

1996: An e-mail message to the "Office" group

name from an electronic address book. I can even indicate with one name that I want the message to be sent to a group of recipients. I frequently send messages to key employees working on the Microsoft Office project, for instance, so I have an addressee called "Office" in my electronic address book. If I choose that entry, the message goes to everyone concerned. Then I type in a short heading for the message so that the recipients will have an idea of what it's about. Then I type the message itself. When the message is transmitted, my name appears automatically in the "From" space.

An electronic message is often just a sentence or two with no formalities. I might send an electronic message to three or four people, saying nothing more than "Let's cancel the 11:00 A.M. Monday meeting and use the time individually to prepare for Tuesday's presentation. Objections?" A reply to my message, in its entirety, might be as succinct as "Fine." If this exchange seems curt, keep in mind that the average Microsoft employee receives dozens of electronic messages a day. An e-mail message is like a statement or a question at a meeting—one thought or inquiry in an ongoing communication.

Microsoft provides e-mail to our employees for business purposes, but like the office telephone, e-mail ends up getting used for personal and social purposes too. A hiker can reach all the members of the Microsoft Hiking Club to try to find a ride to the mountain. A baseball fan can try to find tickets to a sold-out game. And certainly a few romances around Microsoft have benefited from e-mail. When my wife, Melinda, and I were first going out, we took advantage of e-mail. For some reason—probably because it isn't synchronous, real-time communication—people are less shy about sending e-mail than they are communicating on the phone or in person. This can be a benefit or a problem, depending on the situation.

I spend several hours a day reading and answering e-mail from employees, customers, and Microsoft's partners around the world. Anyone in the company can send me e-mail, and because I'm the only person who reads it, nobody has to worry about protocol in a message to me.

I probably wouldn't have to spend so much time reading and answering e-mail if my e-mail address weren't semipublic. When John Seabrook was writing an article about me for *The New Yorker* magazine, he conducted the interview primarily over e-mail. I enjoyed *The New Yorker* piece when it appeared, but it mentioned my e-mail address. The result was an avalanche of mail ranging from students asking me, in effect, to do their homework assignments, to people asking for money, to mail from a group interested in whales who've added my e-mail name to their list. My e-mail address is also a target for both friendly and rude messages from strangers—and provocative ones from the press. ("If you don't answer this by tomorrow, I will publish a story about you and that topless waitress!")

Since I've started writing a syndicated newspaper column, I've had a special e-mail address (askbill@microsoft.com) for questions from readers. It's impossible for me to answer all the mail I get at this address, but I review it. I especially enjoy messages from kids. Here's one of my favorites: "I'm only 13.5 years old, and I want to say I have a problem. My father keeps bugging me. Don't do that, don't do that. Look at Bill Gates. He didn't make all that money watching stupid TV shows. So all I wanted to say is thank you because of you he keeps bugging me. Good bye."

We have other special-subject e-mail addresses at Microsoft—for job applicants, product feedback, and other communications that need to come into the company. Some e-mail on these topics still comes directly to me, and I have to forward it. And three e-mail equivalents of chain letters keep making the rounds. One threatens general bad luck if it isn't forwarded. Another says that your sex life will suffer if you don't forward it. A third, which has been going around since at least 1986 and possibly longer, contains a cookie recipe and a story about a company's having overcharged a woman for the recipe. Supposedly, the woman was told the recipe would cost her "two-fifty," and when her credit card statement came it turned out to be $250. As revenge against the company, according to the mail, she wants you to distribute the recipe for free. In the various versions of this one, different giant food companies or department stores are named. It's the idea of getting back at a big company, any big company, that seems to have made this one such a perennial favorite. When I saw the recipe story the first time, I wondered if it was true. My colleague Nathan Myhrvold remembered seeing the message more than six years earlier but with different specifics. Nathan, whose culinary interests are well known around Microsoft, whipped up a batch of the cookies and concluded that the recipe wasn't worth paying much for in any event.

All of this random "information" gets mixed in with e-mail I really need to see. Fortunately, e-mail software is improving all the time, and it now lets me prioritize mail from senders I designate.

When I travel, I connect my portable computer back into Microsoft's electronic mail system every night to retrieve new messages and send off the ones I've written over the course of the day. Most recipients won't even realize that I'm out of the office. When I'm connected to our corporate network from a remote site, I can also click on a single icon to see how sales are doing, check the status of various projects, or access any other management databases. It's reassuring to me to check my electronic inbox when I'm thousands of miles and a dozen time zones away. Bad news travels very efficiently on e-mail, and if nothing bad is waiting there in my inbox, I know I don't need to worry.

We now use e-mail in all sorts of ways we hadn't anticipated. At the beginning of our annual Microsoft Giving Campaign, for example, employees receive an e-mail message encouraging them to participate and reminding them that Microsoft matches their gifts. The e-mail message contains an electronic pledge card program. When the icon in the message is clicked, the pledge card appears on the employee's screen and he or she can pledge a cash gift or sign up for payroll deductions. If the employee chooses payroll deductions, the information is automatically entered into Microsoft's payroll database. From the electronic form, employees can direct their gifts to their local United Way or to other nonprofit organizations. If they want to, they can further choose to have a United Way donation go to one or more particular groups and click on links to information about those organizations or about volunteering in their communities. From start to finish, the process is completely electronic. As the leader of the company, I can analyze summary information day by day to find out if we're getting good participation or if we need to have a few more rallies to get out the message about how important we think the giving campaign is.

Electronic information sharing within a company goes beyond exchanging e-mail. One of the first ways Microsoft began using information tools internally was to phase out printed computer reports. In many companies, when you go into a top executive's office you see bound books of computer printouts containing monthly financial numbers, dutifully filed away on a shelf. At Microsoft, those numbers are made available only on a computer screen. When someone wants more perspective, he or she can examine the numbers by time period, locale, or almost any other way. When we first put the financial reporting system on-line, people started looking at the numbers in new ways. They began to analyze why our market share in one geographic area was different from our share somewhere else, for instance. And as we all started working with the information more creatively, we discovered errors. Our data processing group apologized. "We're very sorry about these mistakes," they said, "but we've been compiling and distributing these numbers once a month for five years. These same problems were there all along, and no one mentioned them." People hadn't really been working with the print information enough to discover the mistakes, probably because they couldn't examine the information in the creative ways they could when it came on-line.

The flexibility that comes from having the information available electronically is hard to convey to a "nonuser"—a person who doesn't use personal computers in this way. I rarely look at our financial reports on paper anymore. I much prefer to view them electronically.

When the first electronic spreadsheets appeared in 1978, they were a vast improvement over paper and pencil. What they made possible was putting the formulas that produced the data behind each item in a table of data. A particular formula was likely to be tied to other items in the table too, so any change in one value would be immediately reflected in other cells. Projections of sales and growth or changes in interest rates could be played with in "what if" scenarios, and the impact of every change would be instantly apparent.

Current spreadsheets let you view tables of data in different ways. Simple commands enable you to filter and sort the data. The spreadsheet application I know best, Microsoft Excel, includes a feature called a pivot table that allows you to look at summarized information in a multitude of ways. The summarizing criterion can be changed with the click of a mouse. The order of items can be changed by using the mouse to drag a column header from one side of the table to another. It's simple to change from a high-level summary report to an analysis of any data category or an examination of the details one by one. As an individual, you're empowered by decentralized computing power and software. In the old world of mainframes and printed reports, you needed to ask the central MIS or IT department to generate a new report if you wanted to see sales numbers in a different way—and you might have had to wait for a few weeks. With your personal computer and powerful analysis software, you can turn raw data into meaningful information yourself, with no wait.

A pivot table containing sales data by office, product, and sales channel for current and previous fiscal years is distributed electronically to all Microsoft managers every month. Each manager can quickly construct a personal view of the data for his or her requirements. A sales manager might compare sales in his region to budget or the prior year. A product manager can look at her products' sales by country and sales channel. Thousands of possibilities are just a click and a drag away.

But despite the remarkable progress in information sharing inside

	A	B	C	D	E	F
1	Year	1996 ▼				
2	Salesperson	(All) ▼				
3						
4	Sum of Sales	Region				
5	Product	East	North	South	West	Grand Total
6	Gasoline	1,722	8,019	53,160	71,935	134,836
7	Heating Oil	27,498	11,098	4,891	36,670	80,157
8	Lubricants	2,294	1,531	993	3,527	8,345
9	Grand Total	31,514	20,648	59,044	112,132	223,338

Pivot table showing sales data for 1996 summarized by product type and territory

	A	B	C	D	E	F
1	Year	1996 ▼				
2	Salesperson	Adams ▼				
3						
4	Sum of Sales	Region				
5	Product	East	North	South	West	Grand Total
6	Gasoline	1,722	8,019	2,420	15,154	27,315
7	Heating Oil	6,955	11,098	2,516	9,886	30,455
8	Lubricants	-	1,531	436	1,512	3,479
9	Grand Total	8,677	20,648	5,372	26,552	61,249

The same pivot table after one click on the salesperson selector, showing sales data for 1996 for one salesperson by product type and territory

	A	B	C	D	E	F	G	H	I
1									
2									
3	Region	(All) ▼							
4									
5	Sum of Sales	Product	Year						
6		Gasoline		Heating Oil		Lubricants		Grand Total	
7	Salesperson	1995	1996	1995	1996	1995	1996	1995	1996
8	Adams	40,251	27,315	28,804	30,455	3,435	3,479	72,490	61,249
9	Barnes	31,135	56,781	45,045	26,784	622	2,015	76,802	85,580
10	Cooper	40,936	50,740	28,770	22,918	1,475	2,851	71,181	76,509

The same pivot table after the "Product," "Year," and "Salesperson" labels were dragged to new positions, showing sales data for 1995 and 1996 summarized by salesperson and product type companies, one vital function hasn't improved fast enough. Finding the files you want on a corporate network has never been easy because you've had to remember details such as server names, folder names, and filenames—until recently restricted to 8 characters in length, which kept them frustratingly undescriptive. Even people who knew all the right names, or thought they did, would get flummoxed when somebody else changed the name, location, or contents of a file.

What we really need as users is a page that describes what's going on, that answers our questions. Is this the most recent version? Are there similar files I might need? Can I share this file with a customer? Who's the contact for more information? Is there similar information in a file I don't know about? In short, what we want on a corporate network is a Web page for each collection of files, with that Web page giving us information about the files and letting us access them by clicking on their names or on their icons. And we want the ability to use Internet search tools to locate information and to be able to click on links to related information.

I liken setting up an intranet to fitting the final piece into a jigsaw puzzle. Businesses around the world are finding themselves in this happy position now. They've worked hard for years, investing in hardware, software, and training so that they can share information more easily. It's been a sound investment, but not quite all of the pieces have fallen into place yet. Creating documents has been made easy, but finding them has remained hard. Setting up an intranet will complete the picture—and it will cost many companies almost nothing to do because they've already bought all of the expensive pieces.

Intranets are catching on like wildfire in the corporate world, and with good reason. Just as e-mail is replacing many written memos, phone calls, and meetings, the intranet Web page is replacing reports, manuals, forms, filing cabinets, phone directories, cork bulletin boards, and newsletters. Internal Web sites have become repositories of corporate procedures and sources of the latest information on project status, revenues, and decisions of all kinds.

On Microsoft's intranet we have thousands of pages of internal information. A developer wanting to know about project status visits a Web page that provides the summary information and offers links to detailed reports. The developer just clicks on what she wants, confident that she'll get it and not have to waste a lot of time finding it. Our internal Web has all the key information about every product, every division, and every major strategy. Employees can see presentations on hundreds of topics, find the latest information on personnel policies or job openings, or even read a list of Asian holidays that might interrupt business. They can find pricing information, product datasheets, white papers on various technologies, marketing bulletins, the latest documentation style decisions, briefings, and product release information. Our intranet Web pages are a remarkably complete guide to every facet of the company from an inside perspective. They complement the thousand or more Web pages on our public Internet site (http://www.microsoft.com), which provide comprehensive information about the company and its products and initiatives to outsiders.

If I had any doubts about the value of Microsoft's intranet, they were dispelled when I learned that certain kinds of financial data are accessed five times more often now that they can be reached through a Web page. The employees who use the data are sophisticated about computers and

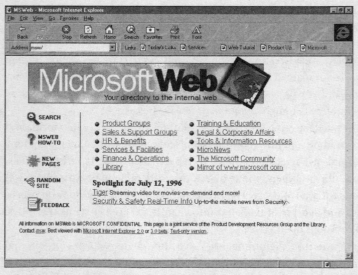

1996: A Web page from Microsoft's intranet

strongly motivated to study the information, and the information wasn't *that* hard to find and use in the first place! The striking increase in usage was a good reminder to me of the benefits of making information self-explanatory and available with just a few mouse clicks—no typing in the name of a server or a file anymore.

To read Web pages, whether on an intranet or the Internet, employees can use Windows, applications they already know, or a "browser" application. Assuming an appropriate corporate network is in place, the only significant expense involved in setting up and maintaining an intranet is the time it takes employees to write the descriptive pages and set up the links—a relatively modest amount of time that's well spent. The tools don't cost a lot. Inexpensive applications that help people create and manage large numbers of linked pages are available from several software companies. Many productivity software applications have built-in capabilities that let employees create Web pages with links, so the same software tools that enable people to write and print memos and spreadsheets also enable them to publish Web pages. These kinds of Internet-related features are at the heart of the updated versions of many office applications. Some people believe that major software applications are already so capable and feature-rich that there will never be a need for newer versions. But there were people who thought the same thing about software five and ten years ago. I have no doubt that new versions of productivity software will continue to be compelling for years to come as speech recognition, social interfaces, and connections to intranets and the Internet are incorporated into core applications.

Personal computing is rapidly evolving toward a Web-based metaphor in which any folder (also called a directory) can be viewed as a Web page. When you look at the contents of a folder, you won't see just filenames or icons—you'll see descriptive text. You can think of a conventional list of files as a Web page so meager that it contains only links to files, with no descriptive text—a Web page that nobody took the time to flesh out with helpful descriptions. A conventional file list is like a page of TV listings that divulges only the times, channels, and names of the shows, not the titles and descriptions of the episodes or the cast or any other informative detail that might help you choose a show to watch.

Microsoft Windows has always let you view a list of files, but it didn't let you design or display a Web page that would show descriptive text and graphics for the list. In the future Windows will display a folder of files either conventionally or as a Web page. We believe there should be no distinction between the way software treats local data, stored on your computer's disk or CD-ROM, and remote data, stored on a corporate network or anywhere on the Internet's World Wide Web. It's all just digital information, and you shouldn't have to use different techniques to navigate it just because it's stored in various places.

As the use of intranets and electronic mail increases, you may find yourself streamlining activities—both internal and external—you didn't even realize were inefficient before. Maybe your business has a tradition of Monday morning status meetings or Thursday afternoon design meetings. Electronic mail and software for collaboration will reduce the need for such meetings. Presentation meetings, called primarily so that participants can listen and learn, can be replaced by Web pages on an intranet or by e-mail messages that have spreadsheets and other exhibits attached. When face-to-face meetings do take place, they'll be more efficient because participants will already have exchanged the background information they need by e-mail.

Companies will use connections to the outside world to make e-mail a convenience for customers too. Think about how a company and a customer have to handle a bill now. The company prints out the bill on a piece of paper and puts it in an envelope that is physically carried to the customer's house. The customer opens the bill, checks his records to see if the amount and the details on the bill seem to be correct, writes a check, and then tries to time mailing the check so that it will arrive shortly before the due date. Businesses and their customers are so used to this process that they don't even think about how wasteful it is. Let's say that the customer disagrees with the bill a company has sent him. He calls up the company and tries to get through to the right person—who might turn out not to be the right person after all, in which case he has to wait for somebody else from the company to come on the line or call him back.

Soon he'll be able to check his PC, wallet PC, or TV set—the information appliance of his choice—for e-mail, including bills. When a bill comes in, he'll be able to check his records, including previous bills, and settle his account electronically on any day he specifies. If he wants to ask about the bill, he'll do that electronically—and asynchronously, at his convenience—by sending the company an e-mail: "Hey, how come this charge is so high?" No more waiting on hold while someone at the company tries to come up with the right person who then has to come up with the answer for him. The appropriate person from the company will call or e-mail him with the answer to his question.

Increased use of e-mail both inside and outside a company will be supplemented by internal and external uses of videoconferencing. Meetings will be conducted electronically by means of shared-screen videoconferencing. Each electronic participant, wherever he or she is, might look at a different physical screen—a video white board, a TV set, or a PC—but each screen will show much the same image. Part of the screen might show someone's face while another part displays a document. If anybody modifies the document, the change will appear almost immediately on all of the screens. In this synchronous sharing, the computer screens will keep up with the people using them. Collaborators at different sites will be able to work together in rich ways.

If a group were to meet electronically to collaborate on a press release, each member would be able to use his or her PC or notebook computer to move paragraphs around and drop in a photograph or a video. The rest of the group would be able to look at the result on their individual screens and see each contributor's work as it actually happened.

We're already accustomed to watching video meetings. Anybody who tunes in to TV news shows such as *Nightline*, which feature long-distance debates, is seeing a videoconference. The host and the guests may be separated by continents, yet they engage in give-and-take as if they were in the same room, and to viewers it almost appears that they are.

Several software packages provide low-cost, low-quality videoconferencing over standard telephone lines. CU-SeeMe, a program developed at Cornell University, lets Windows or Macintosh users communicate video and audio images across the Internet. All that's required other than a PC and a modem is a microphone and a video camera—and you don't need those if all you do is receive signals rather than transmit them.

--------------------------->

Working collaboratively doesn't necessarily require exchanging video pictures. Existing collaboration software enables people at locations all over the world to share applications and documents across the Internet without necessarily videoconferencing. For example, two or more people can work together on a spreadsheet or conduct a meeting in which participants jointly work on text, diagrams, and comments in a shared workspace without video.

But having video is a real plus. To participate in a videoconference that has decent video fidelity you need high-speed connections and better cameras, which often means going to a specially equipped facility—a setting that can be intimidatingly formal. Microsoft has at least one dedicated videoconference room in each of its sales offices around the world. The facilities are heavily used, and they've saved us lots of traveling. Employees in other offices "sit in" on staff meetings, and customers and vendors "visit" us without traveling to our headquarters outside Seattle. Such meetings are becoming popular because they save time and money and are often more productive than audio-only phone conferences or even face-to-face meetings. People seem to be more attentive if they know they're on-camera.

I've noticed that videoconferencing takes some getting used to. If one person is on a videoconference screen, he or she tends to get much more attention than the others in the meeting. I first noticed this when a bunch of us in Seattle were videoconferencing with Steve Ballmer, who was in Europe. It was as if we were all glued to *The Steve Ballmer Show*. When Steve did something like taking off his shoes, we'd all check out each others' reactions. When the meeting was over, I could have told you all about Steve's new haircut but I might not have been able to name the other people who'd been in the room with me. I think this distortion will go away as videoconferences become commonplace.

Some people worry that, by eliminating the subtle human dynamics that go on in a meeting, videoconferences and shared screens will give corporate gatherings all the spontaneity of a congressional photo opportunity. How will people whisper, roll their eyes at a tedious speaker, or pass notes? Clandestine communication will be simpler at a video meeting because the software can allow side communications. Meetings have always had unwritten rules, but when the network is mediating videoconferences, some rules will have to become explicit. Will people be able to signal, publicly or privately, individually or collectively, that they are bored? To what degree should a participant be allowed to block his or her video or audio from the others? Can multiple people speak at the same time? Should private side conversations, one PC to another, be permitted? Over time, as we use videoconferencing facilities, new rules of meeting etiquette will emerge, just as they did with the telephone and e-mail.

When a videoconference has only two participants, it's simply a video phone call. It will be great for saying hello to your kids when you're out of town or showing your veterinarian the way your dog or cat limps. But after the novelty wears off, chances are you'll keep cameras off during most calls you take at home, especially from strangers. A lot of calls will be asymmetric, where the business you're calling provides video to you—of a smiling customer service representative, say—but you simply talk to them. In some situations you may choose to transmit a canned photograph of yourself, your family, or something else you believe expresses your individuality yet protects your visual privacy. Choosing a picture will be akin to choosing a message for your answering machine or voice mail. During the course of a conversation, you might transmit a series of photos of yourself—one smiling, one laughing, one contemplative, and maybe one that's angry—changing them to suit your mood or the point you're making. Live video could be switched on for a friend or when business requires it.

As computers become more powerful, it will be possible for a standard PC to fabricate realistic synthetic images. Your phone or computer will be able to generate a lifelike digital image of your face, showing you listening or even talking. You really will be talking—it's just that you will have taken the call at home while you were dripping wet from the shower. As you talk, your phone will synthesize an image of you in your most businesslike suit. Your facial expressions will match your words—remember, small computers are going to get very powerful. Just as easily, your phone will be able to transmit an image of your words issuing from the mouth of someone else, or from an idealized version of you. If you're talking to someone you've never met and you don't want to show a mole or a flabby chin, your caller won't be able to tell if you really look so much like Cary Grant (or Meg Ryan) or whether you're getting a little help from your computer.

During the years before the broadband infrastructure has been delivered to many homes, businesses will still be able to reach into homes in practical and effective ways. High-quality video won't be involved, but screen images and simultaneous conversation will be shared across midband connections—and even across narrowband connections.

Here's one way it will work: When your company posts information about your products on the Internet, part of that information will be instructions to customers for connecting synchronously with a sales representative who will be able to answer questions through a voice-data connection. Let's say that you operate an outdoors store and a customer is shopping for footwear on your home page, which is really an electronic catalog. The customer might want to know whether the boots she likes are appropriate for climbing. She'll be able to click a button to get your representative to come on the line. The representative will see immediately that the customer is looking at the boots, and he'll have on hand whatever other information the customer has decided to make available about herself—not just her clothing and shoe sizes and style and color preferences, but her athletic interests, her purchases from you in the past, and even her price range. Customers will choose how much information about themselves to make available. Your store's computer may route the customer's inquiry to the same person she spoke to last time, or it may route the customer to someone on your staff who has particular expertise in the product—boots in this case. Without preamble, the customer will be able to ask, "Do these boots work well for climbing?" The representative who answers the question won't have to be in an office. He can be anywhere as long as he has access to a PC and has indicated he's available. If he speaks the right language and has the right expertise, he can help out.

Or let's say you're an attorney and one of your clients decides to change his will. The client calls and you say, "Let's take a quick look at that." You call the will up on your PC, and it appears on your client's screen too. The two of you discuss his needs, and he watches on his screen as you edit the will on your screen.

Another important use of voice-data connections will be to improve product support. Microsoft has thousands of employees whose job is to answer product support questions about Microsoft software. We have as many product support people answering questions about our software as we have engineers building it. The advantage to both us and our customers of all that communication is that we log all the feedback and use it to improve our products. We get lots of customer questions by e-mail, but most of our customers still telephone us. These phone conversations are inefficient. A customer calls in to say that his particular computer is configured in a specific way and is giving him a certain error message. The product support specialist listens to the customer's description and then suggests something, which it takes the caller a few minutes to do. Then the conversation resumes. The average call takes fifteen minutes, and some take an hour. We're looking forward to the product support specialist's being able to see what's on the caller's computer screen and examine the caller's computer directly (with the caller's explicit permission, of course, and carefully so that no one's privacy is invaded). By reducing the need for long descriptions of problems, the new process will reduce the length of the average product support call by 30 or 40 percent, which will make customers happier and cut costs and product prices.

Software operating across the Internet or an intranet will also make it easier to schedule meetings. If your client wants to sit down face-to-face with you to discuss his will, his scheduling program and yours will be able to communicate and pick a date and time you both have free. Then the appointment will just show up on your respective electronic calendars.

This will also be an efficient way to let customers schedule restaurant or theater reservations, but it raises an interesting issue. Let's say that your restaurant isn't getting much business, or that tickets to a show at your theater aren't selling all that well, or that you're a lawyer who doesn't want a client to know that he's your one and only client so far. You might instruct your scheduling program to respond to reservation, ticket, or meeting requests only. The customer's scheduling program wouldn't be able to ask your program to list all the times that reservations are available, or all the theater seating that's available, or all the times your have free for meeting clients. However, if the customer asked for a specific reservation time, or for seats in the orchestra, or for a

----------------------------►

particular two-hour block of time for a meeting, your scheduling program would respond.

Clients will expect their lawyers, dentists, accountants, and other professionals to be able to schedule appointments and exchange documents electronically. A patient might have a quick follow-up question for his doctor—for instance, whether a generic version of a drug is acceptable. Without interrupting, patients and clients will be able to trade e-mail with their doctors and lawyers. We'll see competition based on how effectively one professional group has adopted these communications tools and how much more accessible and efficient this makes them. I'm sure we'll start seeing ads in which firms mention how advanced they are in collaborative uses of the Internet. We see the beginning of this trend already in the increasing number of ads of all kinds that list Web site addresses and e-mail addresses.

All of these electronic innovations—e-mail, shared screens, videoconferencing, and video phone calls—are ways of overcoming physical separation. As they become commonplace, they'll change not just the way we work together but also the distinction we make between the workplace and everywhere else.

By 1996 in the United States there were already more than 8 million "telecommuters" who didn't travel to the office everyday but instead "commuted" via fax machines, telephones, and e-mail. And that was before the Internet's popularity really began to spiral up. Ever-larger numbers of writers, engineers, attorneys, and other people who do their jobs relatively autonomously stay at home for at least a part of their work hours. Salespeople are judged on results. As long as a professional salesperson is productive, it doesn't much matter whether he or she is working in the office, at home, or on the road somewhere. Many people who telecommute find it liberating and convenient, but other people find it claustrophobic to be at home all the time. Still others discover that they don't have the self-discipline to make effective use of the opportunity to avoid a commute. In the years ahead millions more people will telecommute, at least part-time, using the Internet.

Employees who do most of their work on the telephone are strong candidates for telecommuting because calls can be routed to them. Telemarketers, customer service representatives, reservation agents, and product support specialists will have access to as much information on a screen at home as they would on a screen at an office. A decade from now advertisements for many jobs will list how many hours a week of work are expected and how many of those hours, if any, are "inside" hours at a designated location such as an office. Some jobs will require that the employee already have a PC so that he can work at home. Customer service organizations will be able to use part-time labor very easily.

When employees and supervisors are physically separated, management will have to adapt and each individual will have to learn to be a productive employee on his or her own. New feedback mechanisms will have to evolve too, so that both the employer and the employee can determine the quality of the work being done.

An employee in an office is assumed to be working the whole time he or she is there and is paid accordingly, but the same employee working at home might be credited (perhaps at a different rate) only for the time he or she is actually performing work. If the baby starts crying, Dad or Mom will click "Not Available" and take care of the child with unpaid minutes away from the job. When Dad or Mom is ready once again to focus on the job, the "Available" signal will tell the network to start delivering work that needs attention. Part-time work and job sharing will take on new meanings.

The number of offices a company needs might be reduced. A single office or cubicle could serve several people whose inside hours were staggered or irregular. Already the major accounting firms Arthur Andersen and Ernst & Young are among the companies that have replaced large numbers of expensive private offices with a smaller number of generic offices that can be reserved by accountants in from the field. Once the technology is in place, a shared office's computers, phones, and digital white boards could be configured for a particular day's occupant. Or for part of a day an office's white boards could display one employee's calendar, family photos, and favorite cartoons, and later on the same white boards could feature the personal photos or favorite art of a different employee. Wherever a worker logged on, his or her familiar office surroundings could follow, courtesy of digital white boards and the broadband network.

Information technology will affect much more than the physical location and supervision of employees. The very nature of almost every business organization will have to be reexamined. This reexamination should include a critical look at the organization's structure and at the balance between its inside, full-time staff and the outside consultants and firms it works with.

The corporate reengineering movement starts with the premise that there are better ways to design companies. To date most reengineering has focused on moving information around inside a company in new ways. The next movement will be to redefine the boundary between a company and its customers and suppliers. Key questions to reexamine include: How will customers find out about our products? How will customers order? What new competitors will emerge as geography becomes less of a barrier? How can our company do the best job of keeping customers happy after the sale?

Corporate structures will evolve. E-mail and intranets are powerful forces for flattening the hierarchies common to large companies. If communications systems are good enough, companies don't need as many levels of management. Intermediaries in middle management, who once passed information up and down the chain of command, already aren't as important today as they once were. Microsoft was born an Information Age company, and its reporting structure has always been relatively flat. Our goal is to have no more than six levels of management between me and anyone in the company. But in a sense, because of e-mail there are no levels between me and anyone in the company. Unfortunately, all the technology in the world can't give me more hours in a day, so there are many employees I'll never meet or even write to personally. But if they send me e-mail, I read it and make sure they get a response.

As technology makes it easier for a business to find and collaborate with outside expertise, a huge and competitive market for consultants will arise. If you want someone to help design a piece of direct response advertising, you'll ask a software application running on the information highway to list consultants with certain qualifications who are willing to work for no more than a certain rate and who have the right period of time free. Software will do preliminary reference checks for you and help you filter out people who aren't qualified. You'll be able to ask, "Have any of these candidates worked for us before and gotten a rating above eight?" Or, "Will this person agree not to do work for our competitors for a period of time?" This system will become so inexpensive to use that you'll eventually rely on it to find baby-sitters and people to cut your lawn. If you're looking for work as an employee or a contractor, the system will match you with potential employers and be able to send your résumé out electronically with the click of a button.

Companies will reevaluate such employment issues as how extensive a legal or finance department they should maintain, based on the relative benefits of having expertise inside an organization versus a consultant outside it. For particularly busy periods a company will be able to get more help easily, without adding more employees and having to come up with the associated office space. Businesses that successfully draw on the resources available across the network will be more efficient, which will challenge competing businesses to do the same.

Lots of companies will eventually be far smaller because using the Internet will make it easy to find and work with outside resources. Big is not necessarily good when it comes to business. Hollywood studios are surprisingly small in terms of permanent employees because they contract for services—including actors and often facilities—on a movie-by-movie basis. Some software companies follow a similar model, hiring programmers as needed. Of course, companies will still reserve many functions for full-time employees. It would be immensely inefficient to train a succession of outside professionals for a recurring work assignment that could be handled by an employee. Companies will pick areas of "core competency" and use inside employees in those assignments. But a number of functions will be dispersed, both structurally and geographically.

Geographic dispersion will affect much more than corporate structure. Many of today's major social problems have arisen because the population has been crowded into urban areas. The drawbacks of city life are obvious and substantial—traffic, high cost of living, crime, and limited access to the outdoors, among others. The advantages of city life include access to work, services, education, entertainment, and friends.

Over the past hundred years most of the population of the industrialized world has chosen to live in urban areas, after consciously or unconsciously weighing the pluses and minuses.

The broadband interactive network will help change the balance between the advantages and disadvantages. For people who are connected up to it, the information superhighway will substantially reduce the drawbacks of living outside a big city. As a consultant or an employee involved in a service-related field, you'll be able to collaborate with clients or other employees easily from virtually anywhere. As a professional, you'll be able to dispense advice—financial, legal, even some medical— without leaving your house or expecting the client or patient to leave hers. Flexibility is going to be increasingly important as everybody tries to balance family life with work life. You won't always have to travel to see friends and family or to play games. Cultural attractions will be available via the network—although I'm not suggesting that you'll have the same experience watching a Broadway or West End musical in your living room as you would in a New York or London theater. However, improvements in screen sizes and resolutions will enhance all video, including movies, in the home. Educational programming will be extensive. All of this will liberate people who would like to abandon city living.

The opening of the U.S. interstate highway system had a substantial effect on where people chose to settle. It made new suburbs accessible and contributed to the culture of the automobile. There will be significant implications for city planners, real estate developers, and school districts as the opening of the information highway also encourages people to move away from city centers. As large pools of talent disperse, companies will feel even more pressure to be creative about how to work with consultants and employees who aren't located near their operations. This new call for flexibility will set off a positive-feedback cycle, encouraging rural living.

If the population of a city were reduced by even 10 percent, the result would be a major difference in property values and in wear and tear on transportation and other urban systems. If the average office worker in any major city stayed home one or two days a week, the decreases in gas consumption, air pollution, and traffic congestion would be significant. The net effect on city finances is hard to predict. If the people who moved out of cities were mostly the affluent knowledge workers, the urban tax base would be reduced. This would aggravate the inner city's woes and encourage other affluent people to leave. But at the same time, the urban infrastructure might be less heavily loaded. Rents would fall, reducing the cost of living for people who remained in the cities.

It will take decades before the full impact of rich electronic communication is understood. Older people grew up before these communications options existed, and many of them will be reluctant to change familiar patterns. But new generations will bring new perspectives. Our children will grow up comfortable with the idea of working with information tools across distances. These tools will be as natural to them as a telephone or a ballpoint pen is to us. Within the next ten years we'll start to see shifts in how and where we work, the companies we work for, and the places we choose to live. My advice is to get to know these empowering new technologies as best you can. The more you know about them, the less disconcerting—and the more helpful—they will be. Technology's role is to provide flexibility and efficiency, and your goal should be to take advantage of it.

8

FRICTION-FREE CAPITALISM

When Adam Smith described the concept of markets in *Wealth of Nations* in 1776, he theorized that if every buyer knew every seller's price and every seller knew what every buyer was willing to pay, everyone in the "market" would be able to make fully informed decisions and society's resources would be distributed efficiently. To date we haven't achieved Smith's ideal because would-be buyers and would-be sellers hardly ever have complete information.

Not many consumers looking to buy a car stereo have the time or the patience to canvass every dealer, and most consumers thus act on imperfect and limited information. If you end up buying a car stereo for $500 and see it advertised in the paper for $300 a week or two later, you feel foolish for overpaying. But you feel a lot worse if you end up in the wrong job because you haven't done thorough-enough research.

A few markets are already working fairly close to Smith's ideal. Investors buying and selling currency and certain other commodities participate in efficient electronic markets that provide nearly complete instantaneous information about worldwide supply, demand, and prices.

Everybody gets pretty much the same deal because news about all offers, bids, and transactions speeds across wires to trading desks everywhere. Most marketplaces are very inefficient, though. If you're trying to find a doctor, a lawyer, an accountant, or a similar professional or you're shopping for a house, your information is incomplete and it's hard to make comparisons.

The Internet will extend the electronic marketplace and become the ultimate go-between, the universal middleman. Often the only humans involved in a transaction will be the actual buyer and seller. All the goods for sale in the world will be available for you to examine, compare, and often, customize. When you want to buy something, you'll be able to tell your computer to find it for you at the best price offered by any acceptable source or to ask your computer to "haggle" with the computers of various sellers. Information about vendors and their products and services will be available to any computer connected to the network. Servers distributed worldwide will accept bids, resolve offers into completed transactions, control authentication and security, and handle all the other functions of a marketplace, including the transfer of funds. We'll find ourselves in a new world of low-friction, low-overhead capitalism, in which market information will be plentiful and transaction costs low. It will be a shopper's heaven.

Every market, whether it's a bazaar or a broadband network of the future, facilitates competitive pricing and enables goods to move from seller to buyer efficiently with minimal friction—thanks to the market makers, whose job is to bring buyers and sellers together. As the interactive network assumes the role of market maker in category after category of goods and services, traditional middlemen will have to contribute real value to a transaction to justify their commissions. Stores and services that until now have profited just because they are "there"—in a particular geographic location, for example—may find that just being there is no longer enough. But the middlemen who provide added value will not only survive, they will thrive, because the information highway will let them make their services available to customers everywhere.

These ideas will alarm some retailers and consumers. Change can feel threatening. But as with so many changes, I think once we get used to them we'll wonder how we got along before them. The consumer will get

-------------------------->

not only competitive cost savings but also a much wider variety of products and services to choose from. We might end up with fewer stores, but as many stores will be open for business as consumer demand can support. You won't have to go to a store, though. Using the Internet to buy a gift for a friend, you'll be able to consider more choices and you'll often find something more imaginative than you would on a trip to the usual stores. You'll be able to use the time you save to think up a fun clue to put on the package or to put together a personalized card. Or you could spend the time you save with the friend.

Today's Internet doesn't begin to offer the abundance of commerce-related services consumers will enjoy on tomorrow's broadband interactive network, but the Internet is already being used for shopping. On-line catalogs are beginning to appear, and people who take the time to browse discover good buys as well as unusual shopping opportunities. A colleague of mine who wanted the first edition of this book in Italian was able to order it by finding the Web page of a bookstore in Italy and inquiring via e-mail. A helpful store employee responded, in English, and happily took his credit card number.

We all recognize the value of a knowledgeable salesperson when we're shopping for insurance, clothes, investments, jewelry, a camera, an appliance, or a home. We also know that the salesperson's advice is sometimes biased because he or she is ultimately hoping to make a sale from a particular inventory.

On the network, lots of product information will be available directly from manufacturers. As they do today, vendors will use a variety of entertaining and provocative techniques to attract us. Advertising will evolve into a hybrid, a combination of today's television commercials and infomercials, magazine ads, and detailed sales brochures. If an ad catches your attention, you'll be able to ask for more information directly and very easily. Links will help you navigate through whatever information the advertiser has made available. Product literature and manuals might feature video, audio, and text. Vendors will make getting information about their products as simple as possible.

You'll get more than product feature information. The on-line world empowers you to bring pricing anomalies into clear view—great news if you're a consumer, but probably not if you're a high-priced producer or high-markup retailer. The effects of giving full information to the buyer are being felt already, a point that came home to me when the word "Internet" caught my eye in a headline at the top of the front page of the *Financial Times* in the spring of 1996. The story began, "German exporters, battling against a strong currency and high labour costs, have found another cause for their declining share of international markets: the Internet." The article quoted a German trade association official who said that niche export markets were being lost as foreign buyers used the worldwide network to find lower prices. "Where once a German company would offer to supply goods abroad at a given price and be fairly sure of winning the order," the story said, "it was now likely to find the potential customer quoting more competitive prices from perhaps five other suppliers and putting the German company under pressure to improve its terms. The information used by a potential customer with such devastating effect has been garnered by surfing the Internet."

At Microsoft we're using the Internet to get information about our products out. For years we've printed millions of pages of product brochures and data sheets and mailed them out to people who ask for them. But we've never known how much information to put onto a data sheet. We haven't wanted to intimidate casual inquirers, but we know there are people out there who want to know all the product specification details. And since the information changes fairly rapidly, we've often been in the position of having to throw out tens of thousands of copies of a brochure because they described a product we were about to update. We expect that a high percentage of our information dissemination will shift to electronic publishing, particularly because we serve computer users. We've already eliminated the printing of millions of pages of paper by sending out quarterly CD-ROMs and using on-line services to reach professional software developers, some of Microsoft's most sophisticated customers. Microsoft's home page on the World Wide Web has quickly evolved into a gateway to detailed information about every Microsoft product and service. It used to be that visiting a company's headquarters was a good way to learn about it, and there is still a place in business for that kind of interaction. But in the case of Microsoft and many other companies, the fastest way to get a good understanding of what we're up to is to delve into our electronic documents on the Web.

The depth of detail and the number of sources of information on the network will increase dramatically. You won't have to depend only on what Microsoft or any other manufacturer tells you about a product. You'll be able to read independent reviews and information. After you've seen a product's reviews, advertising, and multimedia manuals, you might ask if data on the product from government regulatory agencies is available. You could also check to see whether the vendor has surveyed owners and what the results were. Then you might dig deeper into one area of particular interest to you—durability, for instance. Or you could seek out the advice of sales consultants, human or electronic, who create and publish specialized reviews for all kinds of products, from drill bits to ballet slippers. Of course you'd still ask people you know for recommendations too, but you could do it efficiently, by electronic mail.

If you're thinking of doing business with a company or buying a product, you'll be able to check out what customers have to say about it. If you want to buy a refrigerator, you'll look for Web sites and electronic bulletin boards that contain formal and informal reviews of refrigerators and their manufacturers and retailers. You'll get into the habit of checking on-line before you make any significant purchase. When you have a compliment or a complaint about a record club, a doctor, or even a computer chip, it will be easy to find the places on the network where that company or product is discussed and add your opinion. Ultimately this kind of interactive exchange among consumers will mean that companies who don't serve their customers well will see their reputations and their sales decline. Those who do a great job will attract sizable followings through this new form of word-of-mouth.

It's not just the quantity of information that makes a market efficient or inefficient; it's the quality of the information too. To get good information, you'll have to sift with some discernment through the various endorsements and especially the negative comments you find online. Some of the participants might be motivated more by a chance to sound off than by a genuine impulse to share pertinent information. That's certainly true of today's Internet. Although a network etiquette, or "netiquette," is evolving, the network culture is new enough that it's sometimes as wide open as the lawless West of frontier days. This has its advantages, but its drawbacks too.

Let's say that a company manufactures an air conditioner that 99.9 percent of its customers are very happy with. One angry consumer in the remaining 0.1 percent can post insults about the air conditioner, the company that manufactures it, and even individuals in the company to a consumer Web site or bulletin board and then keep sending the messages over and over and over. The effect can be compared to sitting in a big meeting where everybody has a bullhorn with a volume control and the normal level of conversation is, say, 3. Then one or two people decide to crank their volume up to 100 and start shouting. If I happen to look in on the consumer site because I'm shopping for an air conditioner, my visit may be a waste of time because all I'll find there is the shouting. That's unfair to me, to other people shopping for an air conditioner, and to the company manufacturing the air conditioner.

I've seen single-minded voices rant on bulletin boards, sometimes incessantly, about products, companies, groups, and people they dislike. I've seen forum participants subjected to scurrilous insults. The ease with which an individual, any individual, can share his opinions with the members of a huge electronic community is unprecedented. And because the electronic community is so efficient, a fanatic can easily post a piece of hate mail on twenty sites. Some newsgroup discussions collapse into foolish shouting matches. When that happens, other participants in the discussion usually don't know what to do. Some people yell back. A few people try to restore reason. If the shrill exchanges continue, the forum's sense of community is destroyed.

The Internet has relied on peer pressure for regulation. If somebody in a discussion group posts an irrelevant comment, or worse, tries to sell something in an electronic forum whose participants see it as a noncommercial setting, the would-be digressor or merchant may get a withering barrage of insults. The enforcement of net mores has so far been mostly by self-appointed vigilantes who "flame" participants they believe have crossed the line into antisocial behavior.

The commercial on-line services employ volunteers and professional moderators to monitor conduct on their bulletin boards. Forum moderators can filter out some antisocial behavior by refusing to allow insults or copyrighted information to remain on a system's servers. Most Internet forums remain unmoderated, however. Anything goes, and because people can post messages and information anonymously, there's little accountability. We need a more sophisticated process to gather consensus consumer opinions without depending on the Attorney General's Consumer Complaints Division to act as a filter. We'll have to find some way to protect freedom of expression while getting people to turn the volume down so that the net doesn't remain an amplifier for libel or slander or an outlet for mere venting.

The forums I've been discussing are the free, public ones, but there will also be places on-line where professional information and advice will be offered for a fee. You might need an expert to help you sort out all the information that will be available for the same reasons you might need an expert now. *Consumer Reports* magazine offers objective evaluations of a lot of consumer data, but the reviews are aimed at a broad audience—they don't necessarily speak to your particular requirements. Sometimes you need customized advice. If you can't find exactly the information you need on the network, you'll be able to hire a knowledgeable sales consultant, for five minutes or an afternoon. Via videoconference, she'll help you choose products, which your computer will then buy for you from the cheapest reliable source.

I expect the traditional marriage of advice and sales to be much less prevalent than it is now. Although today's advice appears to the customer to be free, the stores and services that offer it have to pay for it, and they add the cost to the price of the goods. Stores that have been charging more for their products because they offer advice from knowledgeable salespeople will have increasing difficulty competing with the discounters who will be operating over the interactive network. On the network, you'll still find some modest price variations in products from one outlet to another. Differences in outlets' return policies, delivery times, and customer assistance will continue to govern pricing to at least some extent.

Some merchants will offer "consultants," and that cost will be added to the sales price, but for important purchases you're likely to want a truly independent guide. The cost of the independent consultation will be offset by the quality of the advice you receive and by the time you save and the lower price you might end up paying at an outlet the consultant guides you to. The prices consultants charge are likely to be very competitive. Suppose you use a consulting service on the network to get information about where to buy an expensive car at the best price and then you buy it. The consultant who has acted as the middleman in the transaction might charge you at a low hourly rate or charge you a small percentage of the purchase price. The rate or percentage will depend on the uniqueness of the service. Electronic competition will have a lot to do with determining the fee.

Over time more advice will be offered by software applications that have been programmed to analyze your requirements and make appropriate suggestions. A number of large banks have already developed "expert" computer systems to analyze routine loan and credit applications, with great success. As software agents become common and voice simulation and recognition software improves, it will begin to feel as though you're talking to a real person when you consult a multimedia document with an agent personality. You'll be able to interrupt, ask for more detail, or ask to have an explanation repeated. The experience will be like chatting with a personable expert.

Today's home shopping television networks are a step toward the discount electronic commerce coming up. In 1994 they sold nearly $3 billion worth of goods despite the fact that they're synchronous, which means that you may have to sit through pitches for countless other items until they offer one you're interested in. On a broadband network you'll be able to browse globally and at your own pace among goods and services. If you're looking for a sweater, you'll choose a basic style and see as many variations as you like, in every price range. You might watch a fashion show or a product demonstration. Interactivity will combine convenience with entertainment.

Today branded products often appear in feature films and television programs. A character who once would have ordered a beer now asks for a Budweiser. In the 1993 movie *Demolition Man*, Taco Bell restaurants seem to be the only fast-food survivors. Taco Bell's corporate parent, PepsiCo, paid for the privilege. (Outside the U.S. there are very few Taco Bells, so PepsiCo paid to have foreign releases of the movie contain Pizza Hut scenes instead. The necessary changes to the movie were made digitally, with no reshooting.) Microsoft paid a fee to have Arnold Schwarzenegger discover the Arabic version of Windows running on a computer screen during *True Lies,* and Apple paid to have its Power-books used in *Mission Impossible.* In the future companies may pay not only to have their products on-screen but also to make them available for you to buy. In an unobtrusive way, the Internet will offer you the option to inquire about images you see. If you're watching a video of *Top Gun* and think that Tom Cruise's aviator sunglasses look really cool, you'll be able to pause the movie and learn about the glasses or even buy them on the spot—if the film has been tagged with commercial informa-tion. Or you could mark the scene and return to it later. If a movie has a scene filmed in a resort hotel, you'll be able to find out where it's located, check room rates, and make reservations. If the movie's star carries a handsome leather briefcase or a handbag, the network will let you browse the manufacturer's entire line of leather goods and either order something or be directed to a conveniently located retailer.

Because broadband networks will carry high-quality video, you'll often be able to see exactly what you've ordered. This will help prevent the sort of mistake my grandmother once made. I was at summer camp, and she ordered some lemon drops to be sent to me. She ordered one hundred, thinking I would get one hundred pieces of candy. Instead I got one hundred bags. I gave them out to everybody and was especially popular until we all developed canker sores.

On the interactive network, you'll be able to take a video tour of that hotel before you make your reservation. You won't have to wonder whether the flowers you sent to your mother by telephone really were as beautiful as you'd hoped. You'll have watched the florist arrange the bouquet. When you're shopping for clothes, everything will be displayed in your size. You'll even be able to see a sweater paired with a jacket you've already bought or you're thinking about buying.

Once you know exactly what you want, you'll be able to get it just the way you want it. Computers will enable the kinds of goods that are mass-produced today to be custom-made for particular customers. Cus-tomization will become an important way for a manufacturer to add value. Increasing numbers of products—from shoes to chairs, from newspapers and magazines to music albums—will be created on the spot to match the exact specifications of a particular person. Often the cus-tomized item will cost no more than a mass-produced one would. In many product categories, "mass customization" will replace mass production, just as a few generations ago mass production largely replaced made-to-order.

Before mass production, everything was made one piece at a time by labor-intensive methods that hampered productivity and kept the standard of living low. Until the first practical sewing machine was built, every shirt was handmade with needle and thread. The average person didn't have many shirts because they were so expensive. In the 1860s, when mass production techniques began to be used to make clothing, machines turned out large quantities of identical shirts, the prices dropped, and even a laborer could afford to own a number of shirts.

Soon computerized shirt-making machines will obey a different set of instructions for every shirt. When you order, you'll indicate your measurements as well as your choices for fabric, fit, collar, and every other variable. The information will be communicated to a manufacturing plant that will produce the shirt for prompt delivery. Delivering goods ordered over the network will become a big business, with intense competition, and as volume grows, delivery will get very inexpensive and fast.

Levi Strauss & Co. is already experimenting with custom-made jeans for women. At a growing number of their outlets, customers pay about $10 extra to have jeans made to their exact specifications—any of 8,448 different combinations of hip, waist, inseam, and rise measurements and styles. The information is relayed from a PC in the store to a Levi's factory in Tennessee, where the denim is cut by computer-driven machines, tagged with bar codes, and then washed and sewn. The finished jeans are sent back to the store where the order was placed, or shipped overnight directly to the customer.

It's conceivable that within a few years everyone will have their measurements registered electronically so that it will be easy to find out how well ready-made clothes will fit or place custom orders. If you give friends and relatives access to this information, they'll find it a lot easier to shop for you. You might have trouble returning custom clothes to a store, though.

Providing customized information is one of the aspirations of the many new media outlets that are springing up on the Web to serve geographic communities and an enormous range of other communities of interest. Many companies hope that instead of paying 60 cents for a newspaper, you'll soon pay them a few cents a day to assemble news for you electronically—and that you'll pay the original publisher of each story selected a little bit too. You would decide how many articles you want to read and how much you want to spend. For your daily dose of news, you might subscribe to several review services and let a software agent or a human agent pick and choose from them to compile your completely customized "newspaper."

A great deal of investment and innovation in the delivery of news lies ahead. Many sports fans are already hooked on ESPNET SportsZone (http://espnet.sportszone.com), a sports-news site developed by Starwave and ESPN, the cable sports network 80 percent–owned by Disney. Microsoft and NBC have formed MSNBC (http://www.msnbc.com), which is both a cable news channel and a presentation of news on the Web. Many newspapers now have Web sites, although so far most of them are simply putting their content intended for print into electronic form rather than taking much advantage of the new opportunities for interactivity, updating, accountability, and efficiency.

Journalism tends to be inefficient today, for both the journalist and the reader. A reporter spends a considerable amount of time restating past events or in other ways providing context for brand new information. A consumer of news often wastes time wading through a restatement of what she already knows, looking for a new development or a fresh detail. Time is also wasted when the consumer doesn't have enough background information to make sense of breaking developments. And it can be frustrating to have to wait for a particular news story, whether in printed form or during a broadcast.

As Web-based journalism evolves, reporters will deliver information about new developments and maintain extensive background information for consumers who want to explore the context of the day's breaking story. Journalists won't have to restate information over and over, and consumers won't have to listen to the restatements. Because news will be delivered interactively, consumers will get the degree of detail and background they want, and they'll get it when they want it—no waiting until after the next commercial or until the next hour.

Important news stories will be both current and under construction at all times. Readers may be able to check a journalist's source material, from news releases to interview transcripts to public documents, which will promote professionalism among news organizations.

All of the potential for electronic efficiency makes some people worry that, if they use interactive networks to shop or get their news, they'll miss out on the serendipity of running into a surprisingly interesting article in the newspaper or finding an unexpected treat at the mall. But these "surprises" are hardly random. Newspaper editors know a lot about their readers' interests from experience. Once in a while the *New York Times* publishes a front-page article about an advance in mathematics. The somewhat specialized information is presented with an angle that makes it interesting to a good number of readers, including some who didn't think they cared about math. In the same way, buyers for stores think about what's new and might intrigue their type of customer. They don't just provide what their customers ask for. The stores fill their window displays with eyecatching new products they hope will lure their customers inside to try new things.

There will be plenty of opportunities for calculated surprise on interactive networks. From time to time your software agent will try to persuade you to fill out a questionnaire about your tastes. The questionnaire might include all sorts of images in an effort to draw subtle reactions out of you. Your agent might make the process fun by giving you feedback on how you compare with other people. The information you furnish will go into a profile of your tastes, which will guide the agent. As you use the system for reading news or shopping, the agent will be able to update your profile. It will keep track of what you've indicated an interest in, as well as of what you've "happened upon" and then pursued. The agent will use this information as it prepares various surprises to attract and hold your attention. Whenever you want something offbeat and appealing, it will be waiting for you. Needless to say, there will be lots of controversy and negotiation about who can get access to your profile information. It will be crucial, of course, that *you* have such access.

Why would you want to create such a profile? In part, so that you can be pleasantly surprised. In part, so that your computer system can do a better job of helping you. I certainly don't want to reveal everything

------------------------------>

about myself, but it would be helpful if an agent knew I wanted to preview any safety features the new model Lexus might have. Or it could alert me to the publication of a new book by Philip Roth, John Irving, Ernest J. Gaines, Donald Knuth, David Halberstam, or any of my other longtime favorite writers. I'd also like it to signal me when a new book appears on a topic that interests me: economics and technology, learning theories, Franklin Delano Roosevelt, and biotechnology, to name a few. I was quite stimulated by a book called *The Language Instinct*, written by MIT's Steven Pinker, and I'd like to know about new books or articles on its ideas. To protect privacy, systems will be established that allow software agents to use profile information without learning anybody's individual identity. For example, an agent could notify me that Halberstam had a new book out, but all that anyone would know is that a list of anonymous people had been notified. My identity—or yours—would be protected.

You'll also be able to surprise and help yourself by following links other people have set up. Lists of recommended sites are popular on the Web, but not many celebrities or other experts offer their own picks. On the Internet a person who has achieved eminence in some field can publish his or her commentary, recommendations, or even worldview in much the same way that successful investors publish newsletters. An Arnold Palmer or a Nancy Lopez might offer individualized golf clinics and links to golf books, articles, and videos they find worthwhile and entertaining. An editor who works at *The Economist* today might start her own service, offering a digest of the news with links to text and video accounts from a variety of sources.

These subscription services, whether human or electronic, will gather information that conforms to your particular philosophy and set of interests. They'll compete with each other on the basis of their talents and their reputations for doing a good job. Magazines fill a similar role today. Many are narrowly focused and serve as customized realities of a sort. A reader who is politically engaged knows that what he or she is reading in the *National Review* is not "the news." It's a bulletin from the world of conservative politics, one in which little of what the conservative reader believes is challenged. At the other end of the political spectrum, *The Nation* knows its readers' liberal views and biases and sets out to confirm and massage them.

Sooner or later almost every company will probably provide a home page to make it easy for customers to find out about its products and services. Any company that already has a successful distribution strategy in place—in Microsoft's case, using software retailers—has to decide how much to exploit this new opportunity. Even Rolls-Royce, which has an extremely exclusive distribution system, will probably have a home page where you can see its latest motorcars and find out where to buy them. Putting up the latest information, including the names of your distributors, will be very easy—but it's also important to protect your retailers.

Retailers have done a very good job for Microsoft, and we like the fact that customers can go into stores and see most of our products and that there are salespeople in the stores who can give them good advice. Microsoft's plan is to continue to sell through retailers, although some of the retailers will have electronic storefronts.

Will an insurance company that has worked effectively through agents decide that it wants customers to buy directly from the central office? Will it let its agents, who used to sell only locally, sell electronically nationwide? Sales channels and territories will be tough to define. And each company will have to make a calculated gamble on the factors it thinks matter most to its success. Competition in the marketplace will show which approach works best in this or that product or service category.

Home pages are an electronic form of advertising. Once broadband interactive networks are common, advertisers will have to be creative to capture restless viewers who will have grown accustomed to watching whatever they want, whenever they want, and to being able to skip through almost any program.

Today advertising subsidizes nearly all of the programs we watch on television and the articles we read in magazines. Advertisers place their messages in the programs and publications that attract the largest audience likely to want what they're selling. Companies that place ads also spend a lot of money trying to make sure their advertising strategies are working. On the net, advertisers will also want some kind of assurance that their messages are reaching their targeted audiences. Advertising doesn't pay if everybody chooses to skip by the ad. The broadband network will offer alternatives. One might be software that lets the customer fast-forward past everything except the advertising, which will play at normal speed. Or the net might offer the viewer the option of seeing grouped commercials. When commercials were aired in groups on French television, that five-minute block was one of the most popular time segments.

Today television viewers are targeted on a cluster basis. Advertisers know that a television newsmagazine tends to attract one kind of viewer and professional wrestling another. They buy television commercial time with audience size and demographics in mind. Ads aimed at kids subsidize children's shows; ads aimed at homemakers subsidize daytime shows; car and beer ads subsidize sports coverage. The broadcast advertiser has only aggregated information about the viewers of a show, based on a statistical sample. Broadcast advertising thus reaches many people who aren't interested in the products.

Magazines, because they can have and often do have a narrow editorial focus, are able to aim their advertising at somewhat more targeted audiences—car enthusiasts, musicians, women interested in fitness, even groups as narrow as teddy bear fans. People buying a teddy bear magazine want to see the ads for teddy bears and their accessories. In fact, people often buy special-interest magazines as much for the advertising as for the articles. Computer magazine readers pore over the ads. Successful fashion magazines are more than half advertising. They give their readers the experience of window-shopping without the walking. The advertiser doesn't know the specific identities of the magazine's readers, but it knows something about the readership in general.

The net will be able to make much finer distinctions among consumers and to deliver each consumer a different stream of advertising. This will benefit all parties: the consumers because ads will be better tailored to their specific interests and therefore more interesting to them; and the producers and on-line publications because they'll be able to sell advertisers focused blocks of viewers and readers. Advertisers will be able to spend their ad dollars more efficiently. Data on preferences will be gathered and disseminated without violating anyone's privacy because the interactive network will be able to use information about consumers to route advertising without revealing which specific households have been targeted. A restaurant chain will know only that a certain number of middle-income families with small children received their ad.

A middle-aged executive and her husband might see an advertisement for retirement property at the beginning of an episode of *Home Improvement* while the young couple next door might see a family vacation ad at the opening of the same show, regardless of whether the two couples watched the show at the same or a different time. Such finely focused advertisements will be of more value to the advertiser (will result in more sales), so a viewer could conceivably subsidize an entire evening of television by watching a small number of the ads.

Some advertisers—Coca-Cola, for example—want to reach everybody. But even Coca-Cola might decide to direct diet cola ads to households that have expressed an interest in diet books. The Ford Motor Company might want affluent people to be shown a Lincoln Continental ad, young people to see a Ford Escort ad, rural residents to watch an ad for full-size pickup trucks, and everybody else to be sent a Taurus ad. Or a company might advertise the same product for everyone but vary the actors in the ad by gender or race or age. They may revise the ad copy to target particular purchasers. To maximize the value of the advertising, complex algorithms will be required to allocate ad space within a show for each viewer. This will take more effort, but because it will make the messages more effective, the effort will be a good investment.

Even the corner grocery and the local dry cleaner will be able to advertise in ways they never could before. Because individually targeted ad streams will be flowing through the network all the time, video advertising is likely to become cost-effective even for small advertisers. A store's ads might target only a few city blocks and address very specific neighborhoods or communities of interest.

Today the most effective way to reach a highly targeted audience is with a classified ad. Each classification represents a small community of interest: people who want to buy or sell a rug, for example. On the Internet, the classified ad won't be tied to paper or limited to text. If you're looking for a used car, you'll send out a query specifying the price range, model, and features that interest you, and you'll see a list of the available cars that match your criteria. Or you'll ask a software agent to notify you when a suitable car comes on the market. A car seller's ad might include links to a picture or a video of the car or even the car's maintenance records so that you can get a sense of what shape it's in. You'll be able to find out the number of miles on the car the same way, and whether the engine has ever been replaced, and whether the car has air bags. You might want to cross-link to police records, which are public, to see whether the car has been in a wreck.

If you put your house on the market, you'll be able to describe it fully and include photographs, video, floor plans, tax records, utility and repair bills, even a little mood music. The chances that a potential buyer for your house will see your ad are improved because the network will make it easy for anyone to look it up. Many real estate listings are appearing on the Web already, although the amount of information provided about a property tends to be minimal. This will change once there is more competition among suppliers of on-line real estate information. The whole system of real estate agencies and commissions might be changed if the principals have direct access to so much information.

Experimentation in the delivery of advertising is just beginning. A number of U.S. newspapers have banded together to jointly publish their help-wanted classifieds ads electronically, for example (http://www. careerpath.com). This is a pretty limited innovation, but it's handy for people who want to see what kinds of career opportunities are available over large geographical regions.

Every city of any size in the U.S. already has several companies offering classified advertising on the Internet. The first of these services weren't particularly compelling because not many people used them. They were marketplaces without many buyers or sellers. But this state of affairs will change. At some point word-of-mouth from satisfied customers will draw more and more users to the best of the network's classified ad services. A positive-feedback loop will be created as more sellers attract more buyers and more buyers attract more sellers. When a critical mass is achieved, which might be only a year or two after a service is first offered, the classified ad site will be transformed from a curiosity into a primary way private sellers and buyers get together.

Direct response advertising—the junk-mail business—is in for even bigger changes. Today a lot of it really is junk. We cut down a lot of trees in order to send out a lot of mail, much of which gets thrown away unopened. A direct response ad on the net will be an electronic interactive multimedia document rather than a piece of paper. It won't waste natural resources, but there will still have to be some way to make sure you don't get thousands of these almost-free communications a day.

You won't be drowned in a deluge of advertising and unimportant information. You'll use software to filter incoming advertising and other extraneous messages and spend your valuable time looking at the messages that interest you. Most people will block e-mail ads except for those about product areas they're particularly interested in. One way for the advertiser to capture your attention will be to offer you a small amount of money—a nickel or a dollar, say—if you'll look at their ad. When you've watched it, or as you're interacting with it, your electronic account will get credited and the advertiser's electronic account will be debited. In effect, some of the billions of dollars now spent annually on media advertising and on printing and postage for direct mail advertising will instead be divvied up among qualified consumers who agree to watch or read ads sent directly to them as messages.

Network mailings offering this sort of paying advertisement could be extremely effective because they could be carefully targeted. Advertisers will be smart about sending messages worth money only to people who meet appropriate demographics. A company such as Ferrari or Porsche might send $1 messages to car enthusiasts on the chance that seeing a cool new car and hearing the sound of its engine will generate interest. If the ad led to even 1 in 1,000 people's buying a new car as a result, it would be worthwhile to the company. They could adjust the amount they offer depending on the customer's profile. Such ads would be available to consumers not on the advertiser's A-list too. If a sixteen-year-old car-crazy kid wanted to experience the Ferrari ad and was willing to do it for nothing, he could get the message.

Paying people to watch ads might sound a little strange, but it's just another use of the market mechanism for friction-free capitalism. The advertiser decides how much money it's willing to bid for your time, and you decide what your time is worth.

Advertising messages, like the rest of your incoming network mail, will be stored in various folders. You'll instruct your computer how to do the sorting for you. Unread mail from friends and family members might wait for your attention in one folder. Messages and documents that relate to a personal or business interest might be in another folder. And advertisements and messages from people and organizations you don't know could be sorted by how much money was attached to them. You might have a group of 1-cent messages, a group of 10-cent messages, and so forth. You might instruct your computer to refuse all mail that doesn't fit into one of your designated folders and to which no fee is attached. You'd be able to scan each message and dispose of it if it wasn't of interest to you. Some days you might not look into any of the advertising message folders. But if somebody sends you a $10 message, you'll probably take a look—if not for the money, then just to see who thought reaching you was worth $10.

You won't have to take the money someone bids, of course. When you accept a message, you'll be able to cancel the payment. The bid is really just an amount the sender puts at risk to get your attention. (The sender's credit will be checked in advance.) If a man sent you a $100 message saying he thinks he's your long-lost brother, you might forgive him the money if, in fact, he turned out to be your brother. On the other hand, if he was just somebody trying to get your attention to sell you something, you'd probably keep the money, thank you very much.

In the United States, advertisers currently spend more than $20 a month per American family to subsidize free broadcast and cable television. Advertising has become so familiar to us that it doesn't really bother us when we watch television or listen to the radio. We understand that programs are "free" because of the commercials. Customers pay for advertising indirectly because advertising costs are built into the prices of cornflakes, shampoo, and diamonds.

We also pay for entertainment and information directly, when we buy a book or a movie ticket or order a pay-per-view movie. The average American household pays a total of $100 a month for movie tickets, subscriptions to newspapers and magazines, books, cable television fees, tapes and compact discs, video rentals, and the like.

When you pay for entertainment by buying a tape or a CD, your rights to reuse or resell it are restricted. If you buy a copy of *Abbey Road* by the Beatles, you're actually buying the physical disc or tape and a license to replay, any number of times, for noncommercial purposes, the music stored on it. If you buy a paperback book, what you're really buying is the paper and ink and the right to read, and to allow others to read, the words printed on that particular paper with that particular ink. You don't own the words, and you can't reprint them except in narrowly defined circumstances. When you watch a television show, you don't own it either. In fact, it took a United States Supreme Court decision to confirm that people in the U.S. can legally videotape a television show for their personal use.

Eventually the network will enable innovations in the way that intellectual property, such as music and software, is licensed. Record companies, or even individual recording artists, might choose to sell music a new way. You, the consumer, won't need compact discs, tapes, or any other kinds of physical media. The music will be stored as bits of information on a server on the net. "Buying" a song or an album will really mean buying the right to access the appropriate bits. You'll be able to listen at home, at work, or on vacation without carrying around a collection of titles. Anyplace you go where there are audio speakers connected to the Internet, you'll be able to identify yourself and take advantage of your rights. You won't have license to rent a concert hall and sell tickets to play the recorded music or to create an advertisement that incorporates it. But in any noncommercial setting, anywhere you go, you'll have the right to play the song without additional payment to the copyright holder. In the same way, the net could keep track of whether you'd bought the right to read a particular book or see a movie. If you had, you'd be able to call it up at any time, from any information appliance anywhere.

This personal, lifetime buyout of rights is similar to what we do today when we buy a music disc or a tape or a book—and that sounds comfortingly familiar—except that tomorrow there will be no physical medium involved.

All kinds of entertainment pricing schemes will be tried. A song could be made available on a pay-per-hearing basis. (Remember juke boxes?) Each time you listened to it, your account would be charged some small amount, such as 5 cents. At that rate, it would cost 60 cents

----------------------------->

to listen to a twelve-song "album." You'd have to play the whole album twenty-five times to spend $15, which is roughly what a compact disc sells for today. If you found that you liked only one song on the album, you could play it three hundred times, at a nickel each time, for your $15. And because digital information is so flexible, you won't have to pay for the same music again just because it's been released on a new medium, the way people did when they bought CDs to replace the LPs in their music libraries.

We may see digital entertainment that has an expiration date or that allows only a certain number of plays before it has to be bought again. A record company might offer a song for a very low price but let you play it only ten or twenty times. Or they might let you play a song—or an addictive game—ten times free before asking whether you want to buy it. This kind of "demo" usage might replace part of the function served by radio stations today. A music publisher could allow you to pass a new song along to a friend via net mail, but she'd be able to listen to it only a few times before she'd start to get charged for it. A music group could offer a special price, far lower than the price of buying every album they've ever made individually, for a buyer who wanted all of their work.

Even today, paying for entertainment information isn't without nuances. The limited time value of entertainment information affects the way both publishers and film studios market their products. The book publisher often has two release windows, for hardcover and paperback editions. If a customer wants a book and can comfortably afford it, he or she pays $25 to $30. Or the customer can wait for between six months and two years to buy the same book, in a somewhat less expensive and long-lasting format, for $5 to $20.

Successful movies are shown progressively in first-run theaters, in secondary theaters, in hotel rooms, on pay-per-view TV, and on airplanes. Then they're available as video rentals, on premium channels such as HBO, and eventually on network TV. Still later they appear on local television and basic cable channels. Each new outlet brings the movie to a different audience as customers who missed the previous forms of release (accidentally or on purpose) take advantage of the new opportunity.

On the interactive network, various release windows for content will almost certainly be tried. When a hot movie, multimedia title, or electronic book is released, there may be an initial period during which it's priced at a premium. Some people will be willing to pay a high fee, perhaps as much as $30, to see a movie at the same time it appears in the first-run theaters. After a week, a month, or a season, the price will drop to the $3 or $4 we're charged today for pay-per-view movies. Marketers may try some wild schemes. Perhaps a movie will come along that you won't be able to see at all in its first month of release unless you're one of the top 1,000 bidders in an electronic auction on the highway. At the other extreme, if you have a track record of buying movie posters and merchandise related to what you watch, you may find you can get certain movies for next to nothing or with few, if any, commercial interruptions. The opportunity to promote the purchase of *Toy Story* and *Aladdin* merchandise might justify Disney's allowing every child in the world one free viewing.

The transferability of information will be at the center of another big pricing issue: lending rights. Today the lending of computer software is illegal in the United States unless the software publisher allows it. But the lending of physical books, music CDs, and videotapes is legal. At some point the Internet will enable one person to lend another a particular piece of intellectual property at the speed of light, to the extent that these temporary transfers are permitted. Almost all of the music, writing, or other intellectual property stored on discs or in books sits around unused most of the time. When you're not consuming your particular copy of *Thriller* or *Bonfire of the Vanities,* most likely no one else is, either. Publishers count on this. If the average buyer lent his or her albums and books to friends frequently, fewer would be sold and prices would be higher. If we assume that an album is in use, say, 0.1 percent of the time, "light speed" lending might cut the number of copies sold by a factor of 1,000. Publishers of intellectual property of all kinds are likely to want to restrict this electronic lending, so that users will be allowed to lend a copy out perhaps only ten times a year—or maybe not at all. Lending policies should be set by copyright owners, and the industry will need to develop copy management systems that respond to the new market.

Public libraries will become places where anybody can sit down and use high-quality equipment to get access to the Internet's resources. Library committees might use the budgets that pay for buying books, albums, movies, and subscriptions today to fund the royalties they'll pay for using educational electronic materials. Authors may decide to forfeit some or all of the royalties for the "copies" of their work that are to be used in a library.

Copyright laws will need to be clarified to ensure that they work in an on-line environment. The network will force us to think more explicitly in every field and for every market about what rights users have to intellectual property.

Videos, which tend to be watched only once, will continue to be rented, but probably not from stores. Instead, consumers will shop on the interactive network to find movies and other programs deliverable on demand. Neighborhood video rental stores and music stores will face a dwindling market. Bookstores will continue to stock printed books for a long time, but nonfiction and especially reference material will probably be used much more often in electronic than print form.

Everything I've described so far is a direct or indirect implication of an extremely efficient marketplace. Almost any person or business that serves as a middleman will feel the heat of electronic competition.

A small-town lawyer will probably face new competition when legal services are available by videoconference over the network. A person buying a piece of property might choose to consult with a sharp real estate attorney from the other side of the county rather than use a local lawyer who is a generalist. There's recompense, though. The resources of the network will enable the local lawyer to retrain and become an expert in any specialty of her choice. And from her small town, she'll be able to compete electronically in this specialty because of her lower overhead. Clients will benefit as well. The prices for executing routine legal tasks, such as the drafting of wills, will be driven down by the efficiency and specialization of the electronic marketplace. The network will also be able to deliver complicated medical, financial, and other video consulting services. These services will be convenient and popular, especially when they require less time than in-person visits. It will be much easier to make an appointment and turn on your television or computer screen for a fifteen-minute meeting than it will be to drive somewhere, park, sit in a waiting room, take care of your business, and then drive back to your home or office.

Videoconferences of all sorts will increasingly become alternatives to having to drive or fly to a meeting. When you do go somewhere, it will be because it's important that a particular meeting be face-to-face, or because something fun requires that you be there physically. Business travel may fall off, but leisure travel will rise because people will be able to take working vacations, knowing they can stay connected to their offices and homes through the network.

The travel industry will change even though the total amount of traveling may stay the same. Travel agents, like all professionals whose services have been to offer specialized access to information, will have to add value in new ways. Travel agents now search for the availability of travel reservations using databases and reference books customers don't have access to. Once they become familiar with the power of the network and all of the information that will be on it, many travelers will prefer to conduct the searches themselves.

Smart, experienced, and creative travel agents will prosper, but they'll specialize and do more than book reservations. Say that you want to visit Africa. You'll be able to find the cheapest tickets to Kenya yourself, so the travel agency will have to be able to provide something else. The agency might specialize in bookings for nothing but trips to East Africa and hence be able to tell you what sites other customers especially liked, or that the Tsavo National Park is closed, or that if you're really interested in seeing herds of zebra, you're better off visiting Tanzania. Some other travel agents may decide to specialize in selling travel to, rather than from, their own cities. An agent in Chicago might offer services across the network to people around the world who want to visit his hometown, rather than sell services to Chicagoans who want to visit other places. Customers wouldn't know the travel agent the way they tend to know their travel agents now, but the travel agent would certainly know Chicago, which might be more important.

Although many of today's newspapers will be around for a long time, the newspaper business will change fundamentally when the consumer has access to the network. In the U.S., daily newspapers depend on local advertising for most of their revenue. In 1950, when television sets were still novelties, national advertising was responsible for 25 percent of the advertising revenue of American newspapers. By 1993 national advertising contributed only 12 percent, in large measure because of competition from television. The number of daily newspapers in the United States has declined dramatically, and the burden of financing those that remain has shifted to local retail and classified advertising. In 1950 only 18 percent of the advertising revenue of daily newspapers in the United States came from classifieds, but by 1993 that had risen to 35 percent and represented billions of dollars. Classifieds don't really work on radio or television.

The network will provide alternative, more efficient ways for individual buyers and sellers to get together. Once the majority of customers in a market use electronic access to shop, even the newspapers' classified revenues will be threatened. This means that much of the newspaper advertising base could be in jeopardy.

It doesn't mean, however, that newspapers will disappear overnight, or that newspaper companies can't continue to be important and profitable players in the delivery of news and advertising. But like all companies that play a middleman or brokering role, to succeed in the electronic world newspaper companies will have to be alert to change and take advantage of their unique qualities.

Banking is another industry destined for change. About 14,000 banks in the United States cater to retail customers. Most people bank with a firm that has a branch office near their home or on their commuting path. Although minor differences in interest rates and services might shift people from one local bank to another, few customers would consider switching to a branch ten miles out of their way. Today moving your bank records is time-consuming.

But when the network makes geography less important, we'll see electronic, on-line banks that have no branches—no bricks, no mortar— and low fees. Some banks, including one in Britain, are already experimenting with serving customers exclusively by telephone. Tomorrow's low-overhead electronic banks will be extremely competitive, and transactions will be made through computer appliances. There will be less need for cash because most purchases will be handled with a wallet PC or an electronic "smart card" that will combine the features of a credit card, an automatic teller machine card, and a checkbook. All of this is coming at a time when the U.S. banking industry is already consolidating and becoming more efficient.

Banks will be particularly vulnerable to "price wars" in which customers will shop for banking on the basis of costs and interest rates alone. If a bank's services are not to become a mere commodity, the bank will have to find ways to maintain its identity and justify its role as a middleman by adding value. The most robust banks will capitalize on the cost-effective opportunities the interactive network offers for creating, marketing, and delivering banking packages to niche segments. They'll use software tools to create visually rich Web pages to make it easy for customers to find out about and use their services. They'll benefit from immediate and comprehensive marketing data that will enable them to intensify their presence over the network by getting closer and more responsive to their customers. Customers will be able to let the banks know in no uncertain terms what they like and don't like, and the banks will be able to turn on a dime to innovate and to fine-tune their services. Online "branches" that are really custom mixes of electronic banking services tailored to the needs of individual customers can recover the individuality local banks had before branches were forced to be all things to all customers.

A customer of these direct banking services will use his personal-finance software as a filing cabinet, his source of records and data for an integrated view of his finances. The software will gather fresh data from several accounts, analyze spending patterns, calculate tax liabilities, view investments in what-if scenarios and against goals, build reports, and furnish the data for spreadsheets and chart creation.

Consumers will be pleased to discover that a lot of the interest-rate differential between large and small deposits will disappear. With the communications available on the network, middlemen will emerge to aggregate small customers efficiently and get them a rate very close to the low rate large depositors are offered. Fees will be generated for all of these services, but the fee structure will be based on broad, efficient competition.

It wasn't so long ago that a small investor who wanted to put his money into anything beyond a passbook savings account was stymied.

-------------------------------- ➤

The world of stocks and beyond—mutual funds, penny stocks, commercial paper, debentures, and other arcane instruments—was simply off-limits to anyone who wasn't a Wall Street insider.

But that was before computers changed things. Today "discount" stockbroker listings are plentiful in the Yellow Pages, and quite a few investors make their stock purchases from a machine at a local bank or over the telephone. Some brokerage houses now offer low-cost stock trading across the Internet at a fraction of what it costs to work directly with a flesh and blood stockbroker. As the network gets more efficient, investment choices will proliferate. Stockbrokers, like other middlemen whose jobs have been merely to chaperon a transaction, will probably have to offer something more. They'll add value by being knowledgeable. Financial services companies will still thrive. The basic economics of the financial industry will change, but the volume of transactions will skyrocket as the network gives the average consumer direct access to financial markets. Investors with relatively small amounts of money to commit will get better advice than they do today and have opportunities to make profits from the sorts of investments now available only to institutions.

When I prognosticate about the changes coming up in an industry, people often wonder if Microsoft plans to go into that field. Microsoft's competence is in building great software products and the information services that go with them. We cooperate with stores, and we sell goods and services ourselves from time to time, as we always have, but we won't become a bank or a conventional store.

Once, when I referred to a bank's back-end databases as technological "dinosaurs," a reporter wrote an article saying that I thought banks themselves were dinosaurs and that we wanted to compete with them. I had to spend more than a year going around the world telling banks I was misquoted and that we're not really planning to set up in the banking business. Microsoft faces plenty of challenges and opportunities in the business we know—whether it's enterprise support, making computer software, making groupware for the Internet servers, or any other part of our business.

Industry after industry will be changed, and change is unsettling. Some middlemen who handle information or product distribution will find that they no longer add value, and they'll change fields, whereas others will rise to the competitive challenge. There is a nearly infinite number of tasks still left undone in services, education, and urban affairs, to say nothing of the growing workforce the Internet itself will require. The new efficiency in the marketplace will create all sorts of exciting employment opportunities. And the broadband network, which will put an immense amount of information at anyone's fingertips, will be an invaluable training tool. A middleman who does decide to change careers and go into computer consulting, say, will have access to the best texts, the greatest lectures, and information about course requirements, exams, and accreditation. There will be dislocations, but overall, society will benefit from the changes.

Capitalism, demonstrably the greatest of the constructed economic systems, has in the past decade clearly proved its advantages over the alternative systems. As the Internet evolves into a broadband, global, interactive network, those advantages will be magnified. Product and service providers will see what buyers want a lot more efficiently than ever before, and consumers will buy more efficiently. I think Adam Smith would be pleased.

9

EDUCATION: THE BEST INVESTMENT

Maybe it's just my innate optimism, but I expect education of all kinds to improve significantly within the next decade. I believe that information technology will empower people of all ages, both inside and outside the classroom, to learn more easily, enjoyably, and successfully than ever before.

The basic purpose of the PC—managing information to support thinking—aligns superbly with the mission of educational institutions. Improving education is the best investment we can make because downstream benefits flow to every part of society. That's why putting computers and the Internet to work in schools is an exhilarating prospect.

As you think about my arguments in this chapter, keep my perspective in mind. I'm not an educator, but I am a learner. And one of the things I like best about my job is that I'm surrounded by other people who love to learn. In a technology business everybody has to acquire knowledge at a prodigious rate. At Microsoft we read, ask questions, explore, go to lectures, compare our notes and findings with each other, consult experts, daydream, brainstorm, formulate and test hypotheses, build models and simulations, communicate what we're learning, and practice new skills. These are the same activities that go on in the best classrooms, but with a critical difference. At Microsoft these learning activities get a boost from the latest computing and communications technologies. Microsoft succeeds because its employees learn efficiently, in part by using information tools.

I've always been concerned about education, but now that I'm a father I find myself giving it even more thought. I've seen from personal experience how learning is enhanced when the right tools are at hand and how difficult it is when good tools and information aren't available. Human potential is wasted when students everywhere—especially children, who just naturally love computers and interaction—don't have access to the information technology that's commonplace in forward-thinking businesses.

Today preschoolers acquainted with cell phones, pagers, and personal computers go into kindergartens where chalkboards and overhead projectors represent the state of the art. The relatively small number of PCs in the schools come from a hodge-podge of sources. They're often incompatible with each other and not powerful enough to run up-to-date software. Quite a few of them don't have the storage capacity or the network connections that would enable them to adequately respond to a learner's curiosity. American schools are littered with obsolete equipment they aren't allowed to sell or give away. Some school computers don't even have hard disks.

It's not just good late-model PCs that are in short supply in schools. Sufficient electrical outlets, conduit for network cables, phone lines for on-line access, secure locations for hardware, and budget for maintenance are missing too. Reed Hundt, chairman of the U.S. Federal Communications Commission, is among the people who have commented on this: "There are thousands of buildings in this country with millions of people in them who have no telephones, no cable television and no reasonable prospect of broadband services," he said. "They are called schools."

The PC's mediocre track record in the classroom is one reason schools have been slow to embrace technology. A lot of people are cynical about technology in education because it has been over-hyped and has failed to deliver on its promises. Teachers have seen "computer labs" come into schools with great fanfare and then fall into disuse. Often teachers don't know exactly what to do with the computers, and they may get little help in figuring it out. A school may make a one-time investment in computers and then not have the funding to maintain and upgrade them, buy software, and make changes in its teaching methods to take advantage of the computers. The rapid rate at which hardware becomes obsolete isn't a good fit with the school purchasing model: Buy once, use forever. When schools buy new PCs, they often buy behind the price/performance curve to save money—which puts them in the position of owning obsolete machines that much faster. PCs are still considered budgetary "extras," so the cost of providing enough hardware to really make a difference seems huge, insurmountable, even though one estimate is that it would take only a 3 percent addition in the total annual expenditure in education to make adequate hardware available.

Although the educational establishment has been less receptive to change than the profit-minded business world has been, few educators now question whether PCs can contribute. The debate has shifted to how and when a PC should be used, how to pay for it, and what new demands using PCs will make on teachers. Some educators remain wary. They don't want to undertake careless experiments with our children's educations. But this conservatism, combined with the exceptional job security many educators enjoy, can foster resistance to the positive opportunities that technology can bring to education.

Given the obstacles to change, skepticism about big improvements in education is understandable. That's why before describing how technology can enhance learning I'm going to argue that the educational system is capable of change and in all likelihood will change—although exactly when is an open question. If we can believe that change really is possible, it's easier to be optimistic about the possibilities for students and teachers alike.

The educational system in the United States, as in many other countries, has been a world apart. Businesses have rushed to adopt productive new approaches, but schools have changed much more reluctantly, when they have changed at all. If someone who dropped out of society twenty years ago walked into an office today, she'd be surprised by the new workplace: fewer layers of management, computers on every desk, electronic mail, fax machines, intranets. But if she walked into an American classroom, she'd find little that was unfamiliar. Two or three aging PCs at the back of the classroom might catch her attention, but that's about it. As she watched the students and the teacher interact, she'd recognize the same roles and patterns she experienced as a child—and that you probably experienced as a child, no matter how old you are.

But if we look back to a watershed time in U.S. history, to another era when business and societal patterns were undergoing fundamental change, we can see that schools do respond.

In the early 1890s mass media exploded. Newspaper readership was growing rapidly, and mass-circulation magazines were just emerging. The world became smaller because of better communications and transportation, and people became acutely aware of how quickly society was evolving and how out of step the educational system seemed to be. Teenagers discovered that messenger jobs, once easy to come by, were suddenly scarce because of the telephone. They realized that they needed new, more complex skills for industry and business, and they increasingly stayed in school. Many of them recognized that a good education could mean a good job. In the United States, secondary enrollment rose quickly, until by 1900 more than one teenager in eleven was in school.

Imagine what education was like then. Tens of millions of immigrants were overwhelming the schools and social services of our big cities. Many of the students could barely speak the national language. They had few skills and little hope. Yet that generation and the next achieved huge increases in their standard of living, in part because the educational system adapted to their needs.

Through most of the nineteenth century the teacher had defined the educational experience almost single-handedly. But as the end of the century approached, people sensed that schools should modernize and prepare their children for success in the new industrial society. Reform movements demanded that formal curricula assume a larger role, and the individual teacher a less dominant role, in deciding what the schools would teach children. There were furious debates that reverberated for decades about whether curricula should emphasize the ability to think or specific knowledge, whether some college or even secondary classes should be electives, whether students should be grouped according to ability, whether natural science had a place in the classroom, and so on. The coming of the industrial era didn't bring about speedy change in the schools, but the schools did change, and over time the changes added up to a restructuring of the education system.

Today, at the end of another century, change is in the air again. People are wondering whether schools are giving their children the skills they'll need to succeed, this time in the Information Age. A new technology revolution is transforming business and putting new demands on our educational system—even while the technology itself is providing the means for meeting these demands. The people who resist change will be confronted by the growing number of people who see that better ways of learning are available thanks to technology. Foremost among the agents of change will be children themselves.

Kids and computers get along just great, partly because kids aren't invested in established ways of doing things. They like to provoke a reaction, and computers are reactive. Parents are sometimes surprised by how taken with computers even their preschoolers are, but the fascination makes sense when you think about how much a young child enjoys interaction—whether playing peek-a-boo with a parent or jabbing at a remote control and watching the channels change.

When my niece was just three years old, she liked to play with *Just Grandma and Me*, a Brøderbund CD-ROM based on a children's book. She had memorized the dialogue in this cartoon story and talked along with the characters. If my niece used the computer's mouse to click on a mailbox, the mailbox opened and a frog jumped out or sometimes a hand appeared and pulled the mailbox door shut. Her ability to influence what she saw on the screen—to answer the question "What happens if I click here?"—kept her curiosity high. The interactivity, combined with the underlying quality of the storyline, kept her involved.

Parents have been complaining lately that a lot of "edutainment" software has too little educational content and too much entertainment content. They've come to expect the PC to play a role in the education of their children. Sales of personal computers into U.S. homes have been extraordinary, and among families with children the penetration rate is approaching 50 percent. In surveys, parents say that "education" and "working at home" are the two main reasons for buying a computer. Moore's Law and fierce competition have combined to make the $1,200 personal computer a fantastic education platform—you get a big hard disk, a color monitor, a CD-ROM drive, audio capabilities, and a modem for connecting to the Internet.

In the U.S., demand for a competitively educated workforce is high. Parents are insecure about their children's employment prospects, having read repeatedly that there will be "two societies" in the future: high-paid knowledge workers and low-paid service workers. Parents who use computers at work and at home will pressure schools to get with the program. Employers are concerned that entry-level workers don't have language skills, mathematical skills, productivity skills, and use-of-technology skills.

Pressure for change isn't coming just from parents and employers. Charter schools, voucher schools, alternative programs within schools, and homeschooling are all growing in popularity. Each of these alternatives gives parents and students more choice. Homeschooling accounts for roughly 1 million K-12 students in the U.S., about 2 percent of the total. About half of that 2 percent are homeschooled at least in part for religious or cultural reasons, but the rest are taught at home by parents who believe they can do a better job than the schools. A main reason the parents say they believe they can succeed is "the computer." Alternative programs within public schools are popular enough that most of them have enrollment caps and waiting lists. By legitimizing diverse models and making it easier to run a nonstandard school, these trends make it more likely that successful technology-intensive public schools will emerge.

At the college level, the echo of the baby boom is starting to drive enrollments up. As the children of baby-boomers graduate from high school, they are surging into colleges, especially community colleges.

Colleges, except perhaps for the elite private four-year colleges, must find ways to educate more students without spending more money. They're looking increasingly at technology-based solutions. Some colleges are offering distance learning courses, typically by televising a campus course and displaying it at a second location with the professor available via telephone to answer students' questions. Other pioneering schools are starting virtual campuses where students learn through the use of video, TV, and computers. In an era of reduced government funding for campus building programs and faculty growth, these trends could force a faster shift to heavy PC use at the core of the college learning experience.

Productivity is the engine that drives improvement in any system. When we want better medical practice, we find ways to deliver treatment less expensively or more effectively or both. When we want cheaper food, we find ways to make agriculture more efficient. Just spending more—"throwing money at the problem," as they say—isn't enough to create widespread improvement. If we want better education—faster learning and better understanding—we have to find ways to get greater results for every dollar we spend.

Most other institutions in society are going to become much more efficient as a result of information technology, and at some point the schools will too. The fit is just too good to ignore: The PC, like education, is devoted to information—how to get it, how to organize it, how to evaluate it, how to use it, how to keep it at hand, how to disseminate it. The productivity advantages of PCs will become so apparent in the years ahead that it will seem unthinkable to exclude students from the benefits.

The transformation won't be sudden. New information technology is providing just incremental improvements so far. And on the face of it, many basic classroom patterns will remain the familiar ones we're used to. Students will continue to go to classes, listen to teachers, ask questions, do individual work and participate in group work, including hands-on experiments, and do homework. But over time, in stages, the proportions will change and the daily habits of students and teachers will change to take advantage of the opportunities the interactive network offers. The many small changes will add up, as they did in the business office over the last two decades, to significant changes in the formal processes of education.

There is widespread interest in having more computers in schools, but the rate at which they are being supplied varies widely. Only a few countries, such as the Netherlands, already have computers in nearly every school. Governments in some countries, including France, have at least pledged to equip all of their schools' classrooms with computers. Britain, Japan, and the People's Republic of China have begun the process of incorporating information technology into their national curricula with a focus on vocational training. Over time—a longer time in less-developed countries—we're likely to see computers installed in almost every classroom in the world. Many children will have PCs at home too. In South Korea more than 25 percent of the machines sold are going into homes, a reflection of the strong family structure in that country, one that puts great emphasis on getting ahead through education.

Making information technology an integral part of the learning experience will provide a number of benefits. I'm going to speculate on several of the advantages and provide a few scenarios, but first I want to offer some reassurance.

Teachers sometimes express the fear that technology will replace teachers. I can say emphatically and unequivocally, *it won't*. Personal computers will not replace or devalue any of the human talent we need for the educational challenges ahead: We need committed teachers, creative administrators, involved parents, and of course, diligent students. Learning with a computer will be a springboard for learning away from the computer. Young children will still need to touch toys and tools with their hands. Seeing chemical reactions on a computer screen can be a good supplement to hands-on work in a chemistry lab, but it can't replace the real experience. Children of all ages need personal interaction with each other and with adults to learn social and interpersonal skills, such as how to work cooperatively.

Some people are afraid that technology will dehumanize schools, isolating students. But anyone who has ever seen a group of kids working together around a computer the way my friends and I first did in 1968, or watched exchanges between students in classrooms separated by oceans, knows that technology can humanize the educational environment. One of the important educational experiences is collaboration, and in some of the world's most creative classrooms, computers and communications networks are facilitating collaboration already.

The computer won't dominate the learning experience, but it will effectively augment learning, especially outside the classroom. Of all the promising capabilities of computers, I'm most excited by their potential to let people pursue their interests easily and as far as they want to go.

Great educators have always known that learning goes on everywhere, not just in classrooms and not only under the supervision of a teacher. The drive to learn by exploration and discovery is deeply rooted in all of us, but even highly motivated learners with excellent teachers haven't always had an easy time of it. Finding information to satisfy curiosity or end confusion is often difficult.

I've always believed that people have more intelligence and curiosity than our information tools have allowed them to develop fully. Most people have experienced the gratifying sense of accomplishment that comes from getting interested in a topic and then finding good information about it. But if a search for information brings you up against a blank wall, you get discouraged. You may begin to think you're never going to understand the subject. If you experience that natural reaction too often, especially if you're a child, your impulse to pursue topics you're curious about will weaken.

I was fortunate. I grew up in a family that encouraged children to ask questions. And I was lucky in my early teens to become friends with Paul Allen. Soon after I'd met Paul, we had a conversation about gasoline. I'd been curious about it, and I asked him where gasoline came from. I wanted to know what it meant to "refine" gasoline. I wanted to know exactly how it was that gasoline could power a car. I'd found a book on the subject, but it was confusing. Gasoline was only one of the many subjects Paul understood, and he explained it to me in a way that made it interesting and understandable. You could say that my curiosity about gasoline fueled our friendship.

Paul had lots of answers to questions I was curious about (and a great collection of science fiction books too). Reciprocally, I was more of a math person than Paul. We decided to learn how to write software together. We became interactive resources for each other: We asked and answered questions, sketched diagrams to illustrate our points, and brought each other's attention to information. We liked challenging and testing each other.

Today's personal computers offer interaction too. Ask a question, get an answer. But the computer is infinitely patient. Ask a thousand questions, get a thousand answers. And when a computer is hooked up to the Internet, there's no end to the answers available to the inquisitive mind. This is a remarkable development because it empowers students of all ages to learn on their own, at their own rates. A teacher will remain essential for most students most of the time, but often the teacher will serve as a guide for students exploring a world of information. I enjoyed school, but I pursued my strongest interests independently. I can only imagine how Internet access to information would have changed my own learning experiences. It's quite likely that the net will move more of the focus of education from the institution to the individual. The ideal of lifelong learning will become a reality for more people.

We all learn best in our own ways. Some people do better studying one subject at a time, while some do better studying three things at once. Some people do best studying in a structured, linear way, while others do best jumping around, "surrounding" a subject rather than traversing it. Some people prefer to learn by manipulating models, and others by reading. PCs can give individuals great latitude to exploit their own best styles of learning and areas of interest, so that the individual isn't penalized for being out of synch with a schedule or the method of an instructor or a textbook. PCs make cross-disciplinary learning easier too. A student can check a historical fact while studying psychology or access a mathematics tutorial when stuck trying to remember a math skill in the midst of a physics lesson.

Cognitive science has shown that PCs can do a better job of supporting varied thinking and learning modes than lecturers and textbooks can. Moreover, by manipulating information in several media, a student can grasp complex processes and concepts more readily. Studies also show that people suffering from attention deficit disorder can "stay tuned" to a computer much longer than they can to a trained therapist. Even students with normal attention spans find that multimedia, interactivity, swift feedback, and a feeling of control make the PC a compelling device for "thought support" when compared to a textbook or a lecture.

Howard Gardner, a professor at the Harvard Graduate School of Education, argues that different children must be taught differently because individuals understand the world in different ways. Mass-produced

------------------------------->

education can't take into account children's differing approaches to the world. Gardner recommends that schools be "filled with apprenticeships, projects, and technologies" so that every kind of learner can be stimulated and accommodated. We're bound to discover all sorts of different approaches to teaching because the network's tools will make it easy to try out new methods and measure their effectiveness.

In chapter 8, I talked about how information technology now enables Levi Strauss & Co. to sell jeans that are at once mass-produced and custom-fitted. Computers and communications create "mass-customized" goods, jeans produced at mass prices with the custom fit that used to call for expensive tailoring. The same dynamic is at work when an Internet site delivers news tailored to your interests. You get custom-fit news at a low cost. Information technology will bring mass customization to learning too. Multimedia documents and easy-to-use authoring tools will enable teachers to mass-customize a curriculum. As with mass-customized blue jeans and electronic newspapers, computers will fine-tune educational materials so that students can follow their own paths in their own learning styles at their own rates.

Regardless of his or her ability or disability, each learner will work at an individual pace—inside or outside the classroom. Workers will be able to keep up-to-date on techniques in their fields. People anywhere will be able to take the best courses taught by the greatest teachers. The net will spread the availability of adult education, including job training and retraining and career-enhancement courses, all over the world. Computers with social interfaces will figure out how to present information so that it's customized for the particular user.

Many educational software programs will have distinct personalities, and the student and the program will get to know each other. A student will ask, maybe out loud, "What caused the American Civil War?" The computer will reply, describing the conflicting theories: that it was primarily a battle over economics or primarily a battle over human rights. The length and the approach of the answer will vary depending on the student and the circumstances. The student will be able to interrupt at any time to ask the computer for more or less detail or ask for a different approach altogether. The computer will know what the student has already read or seen and will point out connections or correlations and offer links to related subjects. If the computer knows that the student likes historical fiction, war stories, folk music, or sports, it will use that knowledge to make the reply more interesting. It will exploit the child's predilections in order to teach a broader curriculum.

Let's say that another teenager wants to find out about gasoline—not in 1970, when I did, but a few years from now. He might not be lucky enough to have a Paul Allen around, but if his school or his local library has a computer linked to rich multimedia information, he'll be able to go as far with the topic as he wants.

Photos, videos, and animations will show him how oil is drilled, transported, and refined. He'll learn the difference between automobile fuel and aviation fuel. If he wants to know the difference between a car's internal combustion engine and a jetliner's turbine engine, all he'll have to do is ask. He'll be able to explore the complex molecular structure of gasoline, a combination of hundreds of distinct hydrocarbons, and to learn about hydrocarbons too. With all the links to more information available to him, his move to satisfy his curiosity on one point can lead him to an infinite number of fascinating topics.

With a few notable exceptions, this kind of educational software and content isn't available yet. There isn't much good curriculum software, and there's only a modest amount of good supplemental software. Huge quantities of information have gone up on the Internet, but not much of it is aimed specifically at students. There really isn't much of a market for educational software yet because schools haven't demanded it.

Nevertheless, some educational software and textbook companies are already delivering computer products that build basic skills interactively in mathematics, languages, economics, biology, and other curricular subjects. Academic Systems of Palo Alto, California, has been working for four years on an interactive multimedia instructional system for colleges that will help teach basic math and English courses. This "mediated learning" approach blends traditional instruction with computer-based learning. Each student takes a placement test that determines his or her level of understanding of a topic and where instruction should focus. The system then creates a personalized lesson plan for the student. Periodic tests monitor the student's progress, and the lesson plan is modified as the student masters concepts. The program reports problems to the

-------------------------------▶

instructor, who can then give the student individual help. The company has had encouraging success with its college algebra course, reporting pass rates 20 to 38 percentage points higher than with traditional methods. The company says that the 7,000 students at sixteen campuses who have taken the course like the new learning materials but that the most successful classes were those in which an instructor was available a good part of the time. These results underscore the point that new technology by itself usually isn't sufficient to improve education but that appropriate technology can almost always help teachers do a better job.

As textbook budgets and parental spending shift to interactive material, thousands of new software companies will work with teachers to create entertainment-quality interactive learning programs. The Lightspan Partnership is already using Hollywood talent to create interactive live-action and animated programs. Lightspan hopes its sophisticated production techniques will capture and retain the interest of young viewers—ages five through eleven—and encourage them to spend more hours learning. Animated characters lead students through lessons that explain basic concepts and then into games that put the concepts to use. The Lightspan lessons are grouped by two-year age spans, and each series is intended to complement elementary school curricula in mathematics, reading, and language arts. The ultimate plan is for these interactive programs to be delivered on broadband networks to homes and community centers as well as classrooms. In the interim, the programs will be offered on CD-ROMs.

CD-ROMs are available today from many sources, and they offer a taste of the interactive experience. The software responds to instructions by presenting information in text, audio, and video forms. Some CD-ROMs are being used in schools and by kids doing their assignments at home, but they have limitations that information delivered across the network doesn't. CD-ROMs can offer either a little information about a broad range of topics, the way a conventional encyclopedia does, or a lot of information about a single topic such as dinosaurs, but the total amount of information available at one time is limited by the capacity of the compact disc. And of course, students can use only the discs they have on hand. Nevertheless, CD-ROM encyclopedias and other interactive research tools are a great advance over paper-based texts for many purposes.

A lot of interactive material is beginning to make the transition from CD-ROM to the Internet's World Wide Web, where capacity is limitless. The Web is already bringing together the work of many great teachers and imaginative scientists, artists, journalists, musicians, authors, and business people. Classroom teachers are learning to draw on this work, and students are beginning to explore it interactively. In time this access will spread educational and personal opportunities even to students who aren't fortunate enough to enjoy the best schools or the strongest family support. Widespread access will encourage every child to make the most of his or her native talents. It won't happen overnight, but I am convinced that it will happen.

Even though there isn't a lot of high-quality educational software yet, some teachers use popular business software to administer their activities and to give students experience with the tools of the modern workplace. At least half of U.S. college students and an increasing number of high schoolers now prepare reports on PCs with word processors instead of using typewriters or writing by hand. Students studying a foreign language can take advantage of the ability of most of the major word

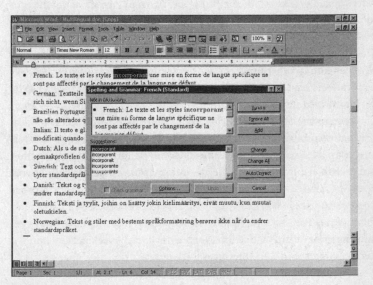

1996: A word processor proofreading a multilingual document

processing programs to work in different languages. Such programs offer supplemental tools for checking spellings and looking up synonyms in a dozen or more languages. Full-blown spreadsheets and charting applications on PCs are routinely used by both students and teachers to explain mathematics and economics theories, and they've become a standard part of most accounting courses.

Enthusiasm for computers and networking is high on campuses. Several universities are centers for advanced research into new computer technologies, and many others maintain large computer labs that students use for collaboration and assignments. Student programmers—who tend to work inexpensively or volunteer their time—are doing impressive work with campus-wide Web sites, and innovative professors keep finding ways to make PC and Internet access more and more valuable to students. At the University of Washington, lesson plans and assignments for some classes are posted on the World Wide Web. Lecture notes are often published on the Web too, a free service I would have loved in my college days.

E-mail over the Internet has aided academic research enormously for years, making it easier for far-flung scientists and other researchers to collaborate. College students everywhere understand the joys of e-mail these days, both for their studies and for keeping in touch inexpensively with their families and friends, including their high school friends who have gone on to other universities. A growing number of parents of college students have become regular e-mail users because that seems to be the best way to stay in touch with their kids.

Some secondary and even elementary schools allow students to have Internet accounts. At Lakeside, my alma mater, the school's network is now connected to the Internet and kids browse for on-line information and exchange e-mail with people all over the world. Nearly all of the students at Lakeside have requested e-mail accounts. In one typical twelve-week period they received a total of 259,587 messages—an average of about thirty messages per student each week. About 49,000 messages came in from outside the school during the twelve-week period, and the students sent about 7,200 messages back out.

Lakeside doesn't know how many e-mail messages each student sends, and it doesn't know what the messages are about. Some of the mail

1996: World Wide Web home page from the Arbor Heights Elementary School

relates to school studies and activities, but undoubtedly a lot of it concerns students' outside interests. Lakeside doesn't view this as an abuse of the electronic mail system but as another way for students to learn.

At Juanita Elementary, a public school not far from Microsoft's headquarters near Seattle, even the youngest students use e-mail. They carry little cards certifying that they've been trained on the Internet and know their "netiquette." They exchange messages with each other and with their teachers, using computers that are spread throughout the school, a few to each classroom. Any student who has parental permission can browse the Internet's World Wide Web or post information on the school's Web page (http://wwwjua.lkwash.wednet.edu/juanita/jes.html)—although students aren't allowed to post pictures of themselves or to identify themselves by more than their initials. The machines in the computer and math labs are so popular during recesses that the school had to abandon a first-come, first-served policy so that the students would stop racing down the hall to get an Internet connection.

Juanita is ahead of most elementary schools in the adoption of technology, but computers and networking are not yet central fixtures of Juanita's curriculum. It's fun to speculate on what educational patterns will emerge in the years ahead as interactive information technology assumes a much larger role in all schools, including elementary schools.

Let's imagine a fifth-grade classroom five years from now, in a school that has a computer for each teacher and one computer for every three students. The computers, which are similar enough that the same software can run on every one of them, are replaced every four years. Our imaginary school also has administrative computers and servers, and all of the school's computers are networked with each other and linked to the Internet on a high-speed connection. As they have for generations, students spend part of each day working alone, part working in groups, and part listening to the teacher and participating in class discussion. But now students studying alone or in groups work with a computer. The computers are in almost constant use, so the teacher rarely has to lead the whole class in discussion. Instead he interacts with students one-on-one or in small groups while the other students use the computers. This results in more personal attention from the teacher for each student. Even during class discussion the smaller group size makes for more eye contact from the teacher and more getting called on.

The computers hold curricular material and support each student's progress through it. Software takes the place of many textbooks, workbooks, tests, and homework exercises. Software also provides much of the content teachers used to deliver by standing at the front of the room and lecturing. The teacher is freed to explain, lead discussion, provide guidance, and motivate. The system tracks each student's progress and keeps students, teachers, and parents informed.

The fifth-grade class devotes a lot of its time to reading and math as well as the other basic subjects such as science and social studies. Sometimes the disciplines are addressed directly, but often they are approached through a "theme." In this approach a variety of academic disciplines are used to study a real-world system such as forest ecology, house construction, criminal justice, or running a small business.

Anyone can quibble with the details. No one's tried to make this scenario come true, in part because the software doesn't exist yet. My scenario is blue sky dreaming, but so were the business scenarios we imagined a decade ago, when we thought that office workers and telecommuters would collaborate electronically via networks and e-mail.

For high school students, educational technology will mean even greater independence. Let's imagine a typical day for Hannah, a fictional eleventh-grader ten years from now.

Hannah begins her morning at home by using a PC connected to the Internet to check her work file on the school server. When she approaches the computer, a tiny video camera attached to the screen feeds the computer a video image of Hannah. The computer recognizes who Hannah is, greets her by name, and—because this is a daily pattern—displays her work file without her having to touch the keyboard or say a word. When she does speak, the computer listens to the sounds of her words and watches her lips move—a combination that allows the PC to be remarkably accurate in recognizing what Hannah is saying. The camera is a standard feature of the PC. Among other things, the PC equipped with a camera lets Hannah and other members of her family videoconference with people outside their home. Of course everybody in Hannah's family is careful about the images they transmit, just as everybody is careful about what they say on a telephone line.

Hannah discovers that her math teacher has made some notes on her homework and given her a new assignment tailored to help her understand the equations she's struggling with. Hannah's "workbook" is on the computer, although she often prints out pages so that she can work out solutions by hand, sometimes on the bus on the way to school.

Later that morning at school, Hannah gives a book report with audio clips of the author's own voice downloaded from the Internet. The class discusses whether or not it's a good idea to find out how an author reads his or her work before you read it for yourself.

Near the end of her lunch period Hannah opens the cafeteria's electronic form on the school's intranet and rates the lunch offerings from most-favorite to least-favorite. It's her way of trying to influence the school's menu for the next few months. While she's on the network, she pulls up a class-planning form and reworks—for the fourth time—a tentative schedule for her senior year. She sends it off with questions to two teachers.

Hannah's incoming e-mail includes a copy of a note from her doctor to her gym teacher giving her a health clearance for gymnastics. It also includes a copy of an exchange her mother has had with Hannah's chemistry teacher, with a request that she try to be less disruptive during individual study time. No more flirting with Jeffrey—except through e-mail, of course.

Hannah spends time in the library researching a paper on the Mexican Revolution using books, CD-ROM reference works, and Internet resources. She also checks in electronically with a scientific expedition in Patagonia that one of her classes is following and reads the log of a researcher who is trying to fix a condensation problem.

When she goes home at the end of the school day, Hannah carries only a couple of books because most of her homework assignments are on the net, ready and waiting for her. Interactive tutorials and other study resources are available too.

Tess, Hannah's older sister, is away at college and has even more control over her own learning. Tess is a sophomore who starts the day in her dorm room by checking the Web for her class schedules and reading assignments.

Tess notices that she's slightly behind in calculus and decides to spend her first couple of hours brushing up. She uses her PC to run through a brief quiz to find out her areas of strength and weakness in the current lesson. As Tess works sets of problems on the screen, she gets instant feedback. She doesn't have to wait a week—or even a minute—for an assignment to be graded and returned. The computer monitors her performance and tests her whenever she wants it to. It also reminds her that she has to get to her Japanese class—in person—in fifteen minutes. Tess sings out "Domo arigato gozaimasu!" and dashes for campus. When she returns early in the afternoon, her computer asks Tess if she wants another calculus quiz or whether she might want to test her written Japanese with a series of on-screen kanji "flashcards."

I'm particularly enthusiastic about the ability computers will give students to quiz themselves. Educators call this the integration of learning and assessment. Tests as we've known them can be pretty depressing experiences for many students, who often associate them with falling short: "I got a bad grade," or "I ran out of time," or "I wasn't ready." After a while many students who haven't done well on tests start to think to themselves, I'd better pretend tests aren't important to me because I'm never going to get good at them. Negative experiences with tests can foster negative attitudes toward education in general.

The interactive network will allow students to quiz themselves anytime without risk. A self-administered quiz is a form of self-exploration, like the tests Paul Allen and I used to give each other. Testing will become a positive part of the learning process. A mistake won't elicit a reprimand; it will trigger the system to help the student overcome a misunderstanding. If someone gets really stuck, the system will offer to explain the circumstances to the teacher. There should be less apprehension about formal tests and fewer surprises; ongoing self-quizzing will give each student a better sense of where he or she stands, make it easier to ask the teacher for help, and—perhaps most important of all—make it easier for the teacher to give the right kind of help.

It always comes back to the indispensable teacher. Unlike the prospects for some professions, the future of teaching looks bright. As innovation has improved the standard of living, there has always been an increase in the part of the workforce dedicated to education. Peter Drucker argues that over the long term teachers will be better compensated as they become more productive, just as other workers have been. He predicts that in the future teaching will be a higher-value, better-paid profession, thanks to technology, and I agree. Educators who bring energy and creativity to a classroom will thrive. So will teachers who build strong relationships with children. Kids love classes taught by adults they know genuinely care about them.

We've all had teachers who made a difference. I had a great chemistry teacher in high school who made his subject immensely interesting. Chemistry seemed enthralling compared to biology, my other science class. In biology, we were dissecting frogs—just hacking them to pieces, actually—and our teacher never explained why. My chemistry teacher prepared exciting lessons and sensationalized his subject a bit, promising that chemistry would help us understand the world. I entered my twenties convinced that chemistry was much more interesting than biology. Then I read James D. Watson's *Molecular Biology of the Gene*, an excellent book that made me realize my high school experience had misled me. Biology, the understanding of life, is now much more interesting to me than chemistry. Biological information is the most important information we can discover because over the next several decades it will revolutionize medicine and lead to treatments for most diseases. Human DNA is like a computer program but far, far more advanced than any software we've ever created.

I wish I had a videotape of some of my chemistry teacher's lectures. A few years ago a friend and I discovered at a university some films of lectures by the distinguished physicist the late Richard Feynman. They were on reels, and we needed a projector. On vacation we were able to watch the lectures many years after Feynman gave the talks at Cornell. We might have gotten more from the lectures if we had been in the lecture hall or if we could have asked Feynman questions via a videoconference. But the clarity of Feynman's thinking brought many of the concepts of physics home to me better than any book I've ever read or any instructor I've ever had. Feynman was a great teacher, and it would be a waste if everybody studying physics didn't have easy access to these lectures and the work of other great teachers. The interactive network will eventually make lots of these valuable resources available to teachers and students—quickly and cheaply.

Unlike Feynman's lectures, my chemistry teacher's great work reached only the few of us who took his course. That's usually been the case. When a teacher prepares wonderful materials to support a particular lesson, only his or her few dozen students benefit each year. And it's practically impossible for even the best teacher to prepare in-depth, interesting material for twenty-five or more students, six hours a day, 180 or more days a year. The Internet will raise productivity in education by enabling teachers to share lessons and materials, building on each other's work so that the best educational materials and practices can spread.

If a teacher in Providence, Rhode Island, happens to have an exceptionally good way of explaining photosynthesis, educators around the world will be able to get her lecture notes and multimedia demonstrations. Some teachers will use material exactly as it comes off the net, but others will take advantage of easy-to-use authoring software that will let them adapt and combine what they find. Feedback from other interested teachers will be simple to get and will help refine a lesson. In a short time the improved material could be in thousands of classrooms all over the world. Teachers everywhere, even in isolated locations, will have access to the best material. It will be easy to tell what work is popular because the network will be able to count the number of times a particular lesson is accessed or poll teachers electronically.

Teachers are discovering that connected computers excel at overcoming cultural isolation too. Computer networks can help their students learn from students in other cultures and participate in discussions all over the world. Teachers in different states and countries are already linking up their classrooms in "learning circles." The purpose of most learning circles is to let students study a specific topic in collaboration with students far away. In 1989, as the Berlin Wall was falling, German students were able to talk about it with their contemporaries in other countries. One learning circle studying the whaling industry included Alaskan Inuit students whose Eskimo villages still depend on whales for food. Students outside the village got so interested they invited an Inuit tribal elder to their class for a learning circle discussion.

Even when computers aren't hooked up to the Internet, teachers are using them to provide compelling interactive lessons for students that let them explore "what if?" scenarios. Several years ago a teacher at Sunnyside High School in Tucson, Arizona, organized a club of students to create computer simulations of real-world behaviors. The students discovered the grim consequences of gang behavior by modeling it for themselves mathematically. The success of the club led to a reorganization of the mathematics curriculum around the idea that education is not about giving the right answer but about giving kids methods to decide whether an answer is right.

Simulation games will get much better, but even now the best of them are fascinating and highly educational. There are already a number of computer models that teach biology. *SimLife*, a popular software program, simulates evolution so that kids get to experience the process instead of just getting facts about it. You don't have to be a child to enjoy this program, which lets you design plants and animals and then watch how they interact and evolve in an ecosystem that you also design. Maxis Software, the publisher of *SimLife*, also produces *SimCity*, which lets you design a city with all of its interrelated systems, such as roads and public transportation. As a player, you get to be the mayor or city planner of a virtual community and challenge yourself to meet your own goals for the community rather than goals artificially imposed by the software's design. You build farms, factories, homes, schools, universities, libraries, museums, zoos, hospitals, prisons, marinas, freeways, bridges, even subways. You cope with urban growth or natural disasters, such as fires. You change the terrain too. When you modify your simulated city by building an airport or raising taxes, the changes can have a predictable or an unexpected effect on the simulated society. Simulation is a great, fast way to better understand how the real world works.

In the future, students of all ages and capabilities will be able to visualize and interact with many kinds of information. A class studying weather will be able to view simulated satellite images based on a model of hypothetical meteorological conditions. Students will pose questions such as "What would happen to the next day's weather if the wind speed increased by 15 MPH?" The computer will model the predicted results, displaying the simulated weather system as it would appear from space.

As simulations become compellingly realistic, we enter the realm of virtual reality. I'm sure that at some point schools will have virtual reality equipment—maybe even VR rooms, the way some schools now have music rooms and theaters—so that students can explore a place, an object, or a subject in this engrossing, interactive way.

Teachers will continue to give homework, but their assignments will include references to electronic resource material. Students will create their own electronic links and use multimedia elements in their homework, which will be submitted electronically across the net. Some multiple-choice homework will be graded by software, and other kinds will be passed along electronically for review by helpers—older students in the same school or even college students many miles away.

What it means to "teach well" will change in some regards, but certainly not in others. Teachers will be pivotal in the future role of educational technology, doing much more than showing kids where to find information on the net. Teachers will still have to understand when to probe, observe, stimulate, or agitate. They'll still have to build kids' skills in written and oral communication. But they'll use technology as a starting point or an aid.

Educators, like so many in today's economy, have to adapt and readapt to changing conditions. First they must make a transition in which some teaching styles and skills will be perceived as more valuable and others less valuable than before. Class sizes may rise by a couple of students in some schools to help pay for the technology and possibly for better teacher compensation. Interacting groups will be smaller, however, and the learning environment more effective, so many teachers may see teaching as a more rewarding profession.

Software will help the teacher as much as the student. Programs will help a teacher summarize information on the skills, progress, interests, and expectations of students. Once teachers have enough information about their students and are relieved of a lot of tedious paperwork, they'll have more time and energy to meet the revealed individual needs of their students. They'll use the information they've gathered to tailor classroom work and homework assignments for individual students. Teachers will be able to keep a cumulative record of a student's work, which can be reviewed any time or shared with other teachers. Teachers and parents will also be able to review and discuss the particulars of a child's progress easily. Add the common availability of video-conferencing, and the potential for strong parent-teacher collaboration will grow.

I expect that teachers will spend a few minutes every week sending e-mail to parents, letting them know in a few words—typed or spoken—how their children are doing. Parents will be able to congratulate their children when they've done something well and will know when their children need help. Parents will be able to follow their children's educational programs much more easily than they can today, when it can be hard to even reach a teacher on the telephone. Assignments, tests, reports, and other information will be available for electronic review. A parent who asks, "What did you do in school today, Honey?" may well know the answer in advance. If a home doesn't have a PC, parents will be able to use the PCs in libraries and other community access points—just as their children will.

One topic that parents, teachers, administrators, and school boards are sure to discuss is the policies that will govern a student's use of information technology. Having students connected directly to limitless

-------------------------->

information and to each other will raise questions not only for parents and schools but for society at large. The issue that comes to most people's minds immediately is whether and how information on the Internet can be restricted or rated, a topic I discuss in chapter 12. There are other questions for the schools, though. Will students be allowed to routinely bring portable PCs with them into every classroom? Will they be allowed to explore independently during group discussions? How much freedom should they have? Should they have access to information their parents find objectionable on moral, social, or political grounds? Should they be allowed to do homework for an unrelated class? Should we prevent them from sending notes to each other during class? Should the teacher be able to monitor every student's screen or record a student's computer activity for spot-checking later? These questions, many of which are merely modern restatements of traditional classroom issues, will be answered differently in different places, often after heated debate that will include the electronic participation of parents and the wider community.

Teachers will use electronic mail to recruit individual parents for on-line classroom forums about topics in which the parents have expertise. Grandparents, professionals, and community leaders will have the opportunity to participate in the teaching process too, even if only for an hour here or there. It will be practical, inexpensive, and commonplace for knowledgeable guests to lead or join discussions over the Internet and eventually in videoconferences from their homes or offices.

Businesses will get increasingly involved too. Their financial participation will be particularly important. Many cable and telephone companies in the United States have already promised free or reduced-price network connections for schools and libraries in their areas. TCI, for example, offers free cable to the schools in the communities it serves, and Time Warner and AT&T have made major commitments. Companies and other organizations are teaming up to get computers, software, and network connections into elementary schools, high schools, colleges, and libraries.

Some businesses have already seen their investments in education make a difference, even for students facing tough challenges. Christopher Columbus Middle School in Union City, New Jersey, was a school created out of crisis. In the late 1980s the state test scores were so low and the absentee and dropout rates for its school district so high that the state considered taking it over. The school system, the teachers, and the parents (more than 90 percent of whom were Hispanic and didn't speak English as a first language) came up with an innovative five-year rescue plan.

Bell Atlantic, the local telephone company, agreed to help fund a special networked, multimedia system of PCs linking the students' homes with the classrooms, teachers, and school administrators. The corporation initially provided 140 multimedia personal computers, enough to supply one to the home of each seventh-grader and each seventh-grade teacher and at least four to each classroom. All the computers were linked to the Internet over high-speed lines, and the teachers were trained to use them. The teachers in turn set up weekend training courses for the parents, more than half of whom attended. Teachers encouraged the students and the parents to use e-mail and the Internet.

Two years later many of the parents were still actively involved with their children's use of the home PCs and used the PCs themselves to keep in touch with teachers and administrators. The dropout rate and absenteeism were both almost zero. The students were scoring nearly three times higher than the average for all New Jersey inner-city schools on standardized tests. It's no surprise that the program has been expanded from the seventh grade now to include the entire middle school.

Raymond W. Smith, chairman of the board and CEO of Bell Atlantic, attributed the success to "a combination of a school system ready for fundamental change in teaching methods, a parent body that was supportive and wanted to be involved, and the careful but intensive integration of technology into both the homes and classrooms." He described the result as "a true learning community in which the home and school reinforce and support each other."

Corporations make these kinds of efforts because they recognize that they have a vested interest in education. I also expect it to become common for companies that want to help out in education to provide recognition and cash awards to teachers whose instructional materials—shared on the Internet, for all to use—are making a difference. In a modest way, this will help provide financial incentives for the spread of best practices.

Many of the educational scenarios I've described so far would be feasible today if schools were better equipped and connected to the Internet, and if the necessary software existed. If we look down the road to the impact of broadband technology, we'll see even more ways to make our schools exciting places to learn.

A classroom will still be a classroom, but technology will transform more of the day-to-day learning experience. Wall-mounted video white boards will replace a teacher's chalkboard handwriting with readable fonts and colorful graphics drawn from the net's rich store of multimedia educational content—millions of illustrations, animations, photographs, and full-motion videos.

I can imagine a middle school science teacher a decade or so from now working on a lecture about the sun that will explain not only the science but also the history of discoveries about the sun. When she wants to show a still picture or a video, whether of a piece of art, a portrait of a solar scientist, or a video simulating the orbiting of the planets around our sun, the net will allow her to select from a comprehensive catalog of images. Snippets of video and narrated animations from countless sources will be available. It will take only minutes to pull together a visual show that would require days of work to organize now. As the teacher lectures about the sun, she can have images and diagrams appear at appropriate times. If a student asks her about the source of the sun's power, she can answer with animated graphics of hydrogen and helium atoms. She'll be able to show solar flares or sunspots or other phenomena, or she might call up a brief video on fusion energy to the white board. The teacher will have organized the links to servers on the global network in advance, and she'll make the list of links available to her students so that during study times in the library or at home they'll be able to review the material from as many perspectives as they find helpful.

Or think of a high school art teacher using a digital white board to display a high-quality digital reproduction of Seurat's *Bathers at Asnières*, which shows young men relaxing on the bank of the Seine River in the 1880s against a background of sailboats and smokestacks. The white board could pronounce the name of the painting in the original French—*Une Baignade à Asnières*—and show a map of the outskirts of Paris with the town of Asnières highlighted. The teacher might use the painting, which presaged Pointillism, to illustrate the end of Impressionism. Or he might use it to get into broader topics—life in France at the end of the nineteenth century, the Industrial Revolution, or even the way the eye sees complementary colors.

He might point to the orangish-red hat of a figure standing on the far right side of the composition and say: "Look at the vibrancy of the hat. Seurat has tricked the eye. The hat is red, but he's added tiny dots of orange and blue. You don't really notice the blue unless you look closely." As the teacher said this, the picture could zoom in on the hat until the texture of the canvas was apparent. At this magnification, specks of blue would be obvious, and the teacher would point out that blue is the complement of orange. A color wheel might appear on the white board, and either the teacher or the multimedia document itself could explain: "Every color on this wheel is arranged opposite its complement. Red is opposite green, yellow is opposite purple, and blue is opposite orange. It is a quirk of the eye that staring at a color creates an afterimage of its complementary color. Seurat used this trick to make the red and orange hues of the hat more vivid. He sneaked in dots of blue."

I'm describing an educational experience that calls for big changes from what we see in schools today, but we know that big change is possible. Just as education in America changed a century ago, it can and will change again. The transformation won't be limited to the United States. It will take different forms in different countries, but information technology will be at the heart of it everywhere. In five years' time we may not see much progress, but within ten years new technology will play a big role in learning both inside and outside the classroom. The return on the investment will be profound.

10

PLUGGED IN
AT HOME

One concern often mentioned in talk about the coming communications revolution is that people won't socialize anymore. Commentators worry that our homes will become such cozy entertainment providers that we'll never leave them, and that safe in our private sanctuaries we'll become isolated. I don't think that's going to happen, and later in this chapter, when I describe the house I'm building, I think I make my case.

The house has been under construction for what seems like most of my life. (It seems like I've been reading about the construction even longer.) Yes, it's full of advanced entertainment equipment—a small movie theater and a video-on-demand system—and it should be an interesting place to live, but I certainly don't plan to stay at home all the time. Neither will other people when they have all kinds of entertainment flowing into their homes. They'll continue to go to theaters, just as they'll continue to visit parks and museums and go shopping. As the behaviorists keep reminding us, we're social animals. It will be practical to stay at home more of the time because the Internet will create so many new possibilities for home-based entertainment and for communications—both personal and professional. But I think people will spend almost as much time out in the wider world as they always have, although the mix of activities we do at home and outside will change.

In chapter 1, I talked about the initial reactions to the railroad and the telephone, the dire predictions that were never realized. More recently, in the 1950s, the pundits said that movie theaters would disappear as everybody stayed at home watching the new invention, television. Later on pay TV and movie video rentals provoked the same kind of talk. Why would anyone spend money for parking and baby-sitters and then buy the most expensive soft drinks and candy bars in the world—all to sit in a dark room with strangers? But popular movies continue to fill theaters. Personally, I love movies and enjoy going out to see them. I go to the movies almost every week, and I don't think the network will change that.

The new communications capabilities will make it far easier than it is today to stay in touch with friends and relatives who live out of town. Most of us have struggled at one time or another to keep alive a friendship with someone far away. I used to date a woman who lived in a different city. We spent a lot of time together on e-mail, and we figured out a way we could sort of go to the movies together. We'd find a film that was playing at about the same time in both of our cities. We'd drive to our respective theaters, talking on our cellular phones. We'd watch the movie, and on the way home we'd use our phones again to discuss the show. In the future, of course, "virtual dating" could be even better because both the movie watching and talk during and after the movie could go on in a videoconference.

I already play bridge on an on-line system that allows the players to see who else is interested in joining a game—they're in what's called a waiting room. The players have a primitive ability to choose the way they appear to the other players: their sex, hairstyle, body build, and so on. The first time I connected up to the system, I was in a rush to keep a bridge appointment so I didn't spend any time setting up my electronic appearance. After my friends and I began playing, they started sending me messages about how I was bald and naked from the waist up, the only part of the body the system showed. Even though this system didn't offer video or voice communication the way systems in the future will, the ability to send text messages to each other while we were playing made it a real blast.

The Internet will not only make it easier to keep up with friends who are far away, but it will also enable us to find new friends. Striking up a friendship across the network will lead naturally to getting together in person. The talk-show host Rush Limbaugh met his wife on an on-line service. Right now our methods for linking up with people we might like are pretty limited, but the network will change that. We'll be meeting some of our new friends in new ways. This alone will make life more interesting.

Suppose you want to find somebody to play bridge with. The network will help you find cardplayers with the right level of skill who are available in your own neighborhood or in other cities or countries. The idea of interactive games played by far-flung participants is hardly new. Chess players have been carrying on games by mail, one move at a time, for generations. Applications running on the network will make it easy to find other people who share your interests. You'll play together at the same pace you would play face-to-face. While you're playing a game— say, chess or bridge or *Starfighter*—you'll be able to chat with the other players across the network.

Playing a friendly group game, as you do at the traditional card table, is pleasurable as much for the fellowship as for the competition. The game is more fun when you're enjoying the conversation too. A number of companies are taking this remote, multiplayer-game concept to a new level. You'll be able to play alone, with a few friends, or with thousands of people, and it will eventually be possible to see the people you're playing with—if they decide to let you see them. It will be easy to locate an expert and watch him play or take lessons from him. On the net, you and your friends will not only be able to gather around a game table, but you'll also be able to "meet" at a virtual representation of a real place—say, Kensington Gardens—or in a completely imaginary setting. You'll be able to play a conventional game in a remarkable location or play a new kind of game in which exploring the virtual setting is part of the action.

The boundary between what is a game and what isn't has been a little unclear at times, and it will get even fuzzier. If you use virtual reality to explore an imaginary world or the inside of the human body, is that a game? If a computer makes learning fun by posing clever questions, is that a game? People tend to think of computer games as action-packed video competitions catering to teenage boys, but that's too narrow a view.

Warren Buffett, a good friend of mine who is famous for his invest-ment savvy, studiously avoided technology and technology investing. For years I kept trying to persuade Warren to use a personal computer. I even offered to fly out to Omaha to get him started. He wasn't the least bit interested until he found out he could play bridge with friends all over the country through an on-line service. That's all it took. For the first six months after that discovery Warren would come home and play bridge for hours on end. Once he tried the computer, he was hooked. Now in many weeks Warren uses on-line services more than I do. The present system doesn't require you to enter your true appearance, your name, your age, or your sex. But most of the users seem to be either kids or retirees (neither of which describes Warren). One feature that had to be added into the system was a limit that permits parents to restrict the amount of time and money their kids spend on-line.

I think on-line computer game playing will catch on in a big way. We'll be able to choose from a wide range of games, including all the classic board and card games as well as action adventure and role-playing games. New styles of games will be invented specifically for the interactive medium. Contests will award prizes. From time to time celebrities and experts will come onto the system, and everybody else will be able to watch the celebrities play or sign up to play against them.

TV game shows will evolve to a new level when viewer feedback becomes an element. Viewers will be able to vote and see the results immediately—sort of like the meter that used to measure live audience applause in old talent shows and *Queen for a Day*—so that prizes can be given to the players. Some entrepreneurial companies, Answer TV for one, have already designed and tested systems specifically for interactive TV games, but because such systems have only one application, so far they haven't caught on enough to make money. On the Internet, you won't have to buy special hardware or software to interact with a TV show. Imagine a future *Password* or *Jeopardy!* show that will let viewers

---------------------------->

at home participate and win either cash or credits of some sort. Shows will even be able to track and reward their regular audience members by giving them special prizes or mentioning them by name if they choose to join the game.

Gambling is going to be another way to play on the Internet. It's a huge business in Las Vegas, Reno, and Atlantic City, and it nearly supports Monaco. The casinos make enormous profits because gamblers continue to believe that even though the odds are against them they're going to win—even at games where it's strictly luck that counts. I've always preferred games of skill, such as poker, bridge, chess, checkers, and Go. Even when luck plays a role, as it does in poker, being able to assess odds is still a valuable skill.

Warren Buffett once offered me a wager that called for quick calculation on my part. He put four dice on a table and proposed that we each choose one, that we discard the remaining two, and that we bet on whose die would roll a higher number most often. He offered to let me choose first.

I knew Warren was up to something because the dice had peculiar combinations of numbers on them. One of the dice, which I'll call Die A, showed the numbers 1, 2, 3, 9, 10 and 11. Die B showed the numbers 0, 1, 7, 8, 8, and 9. Die C showed 5, 5, 6, 6, 7, and 7. Die D showed 3, 4, 4, 5, 11 and 12.

"You get to pick first, so what kind of betting proposition would you like?" Warren asked. "Would you like even odds? Will you give me odds?"

"Let me look at those dice," I said.

After studying the dice for a minute, I said, "This is a losing proposition. You choose first."

Warren chose first, but we both knew I'd won the game. I studied the die he chose and spent a couple of minutes thinking about the remaining three dice before I chose one.

It wasn't immediately evident that because of the clever selection of numbers for the dice they were nontransitive—the mathematical principle of transivity, that if A beats B and B beats C, then A beats C, did not apply. Assuming ties were rerolled, each of the four dice could be beaten by one of the others: Die A would beat B an average of 11 out of every 17 rolls—almost two-thirds of the time. Die B would beat C with the same frequency. Likewise, C would beat D $^{11}/_{17}$ of the time too. And improbable as it sounds, die D would beat A just as often. I didn't figure out all the odds on the spot, of course. I just studied the dice long enough to figure out which one to pick.

Although I play blackjack sometimes when I'm in Las Vegas, I'm looking for a challenge, not to get lucky. That's probably because I'm so much more limited by time than by money. If they had a gambling game that would award the winner a few more hours in the day, I might be drawn in.

Advances in technology have historically had an impact on gambling. One early use of the telegraph and of ticker services later was to deliver racetrack results. Satellite television broadcasts have contributed to off-track betting. Slot machine designs have always tracked the progress in mechanical calculators and more recently the progress in computers. Now the Internet is attracting a lot of gambling attention, although it's cautious attention because what's legal isn't clear yet. The tiny European principality of Liechtenstein even runs a national lottery on the Internet, called InterLotto, with gamblers participating from all over the world. Interactive networks are bound to have a significant effect on both legal and illegal gambling. We're sure to see current odds posted on servers, and people will use e-mail to place bets with electronic currency. Payoffs will be electronic too.

Gambling is a highly regulated industry, and we can't be sure what kinds will be allowed on the net or what the rules will be. Some international airlines are experimenting with letting travelers use video screens to gamble. Perhaps gambling games will have to provide full disclosure of the odds against you. One thing's for sure: The global interactive network will make gambling far more difficult to control than it is today. The technology will allow people to bet on anything they choose to, and if a form of gambling is legal, somebody is sure to set up a service for it. You'll be able to bring horse races, dog races, or any other kind of live sports event into your home in real-time, so some of the excitement of the track or the stadium will be available to you too.

All kinds of people, not just game players and gamblers, will use the interactive network's unique capabilities to find people who have interests in common with them. You may belong to the local ski club so that you can meet other people who like to ski. You may also subscribe to *Recreational Skier* so that you can get information about new ski products. Now, if you frequent the Internet, you can check in with one of the Web sites devoted to discussing and promoting skiing, skiing businesses, and ski areas. You'll find weather reports, "live" photos, and trail reports from most major ski areas. The information is bound to get even richer in the years ahead. The best Web sites devoted to skiing will evolve into communities that will not only provide you with up-to-date information about equipment and weather conditions, but will also offer you a way to stay in touch with other enthusiasts worldwide.

For a typical electronic community, the greater the number of people who join the more valuable it becomes to everybody. Eventually most of the world's skiing enthusiasts will participate in one electronic community or another, at least occasionally. If you join one, you'll find out what the best slopes near Munich are, the lowest price anywhere for a particular set of poles, and the latest news and advertising about all ski-related products. If people have taken photos or made videos of a race or a trip, they'll be able to send them out to everybody. Anybody who has an opinion about a book on skiing will be able to post a review. Laws and safety practices will be debated. Multimedia instructional videos will be available at a moment's notice, free or for a charge, to one person or to hundreds of thousands. This community on the network will be the place to go if you're interested in skiing.

If you want to get yourself in better physical condition before you try a difficult slope, you might find training more fun if you're in close electronic touch with a dozen other people your size, weight, and age who share your specific goals for exercise and losing weight. Members of this community could get together to encourage each other and even work out at the same time. You'd have less to be self-conscious about in an exercise program in which everybody else is like you. If you were still uncomfortable, you could always turn your video camera off.

The community of skiers is quite large and easy to define. But applications on the net will help you find people and information that intersect with any of your interests, no matter how specific. If you're thinking of visiting Berlin, the Internet will make vast amounts of historical, sociological, and tourist information available to you. But applications will help you find fellow enthusiasts there too. You'll be invited to register your interests in databases that can be analyzed by applications that will even find people you might like to meet on your trip. If you have a collection of Venetian glass paperweights, you'd probably enjoy being a member of one or more worldwide communities of people who share your interest. Some of those people may live in Berlin and be delighted to show you their collections. If you'll be taking your ten-year-old daughter with you to Berlin, you might query whether anyone in Berlin who has a ten-year-old and speaks your language would be willing to spend time with you. If you find two or three suitable people, you'll have created a small—even though probably temporary—community of interest.

I recently visited Africa and took a lot of pictures of chimpanzees. If the Internet were more mature and more widely used now, I might put out a message offering to exchange photos with anyone else from the safari, or I might post my chimpanzee photos to a bulletin board that only fellow safari members could access. People are already able to limit access this way because Web server software makes it easy to set up password-protected pages on an Internet site.

Thousands of newsgroups on the Internet and countless forums on the commercial on-line services have been set up as locations where small communities of interest can share information. On the Internet you'll find lively text-based discussion groups with names such as alt.agriculture.fruit, alt.animals.raccoons, alt.asian-movies, alt.coffee, bionet.biology.cardiovascular, soc.religion.islam, and talk.philosophy.misc. But even these topics aren't nearly as specialized as some of the subjects I expect electronic communities will take shape around in the future.

Some on-line communities will be very local, and some will be global. You won't be overwhelmed by the number of choices any more than you are now by the number of entries in a telephone book. You'll look for a group that addresses a general interest of yours, and then you'll search through it for the small segment of it you want to join. I can imagine the administration of every municipality, for example, becoming the focus of an electronic community.

Sometimes I get annoyed by a traffic light near my office that always stays red longer than I think it should. I could write a letter to the city, telling the engineers who program the lights that the timing isn't optimal, but that would be just one cranky letter. If I could find the "community" of people who drive the same route I do, we could send a strong complaint to the city. I'd find my fellow complainers by sending a message to people who work near me or by posting a message on a community affairs bulletin board that showed a map of the intersection accompanied by the message: "During the morning rush hours hardly anybody goes left at this intersection. Does anybody else think the cycle should be shortened?" Anyone who agreed with me could add their agreement to my message.

As on-line communities get more prominent, people will turn to them to find out what the public is really thinking. People like to know what's going on, which movies their friends have been going to, and what news other people think is interesting. I might want to read the same "newspaper front page" as the person I'm going to meet with later today so that we'll have something in common to talk about.

Some institutions will have to make big changes as on-line communities get more powerful. Doctors and medical researchers are already having to contend with patients who explore medical literature electronically and compare notes with other patients who have the same disease. Word of unorthodox or unapproved treatments spreads fast in these on-line communities. Some patients in drug trials have been able to figure out by communicating with other patients in the trial that they are receiving a placebo rather than the real medication. The discovery has prompted some of them to drop out of the trials or to seek alternative, simultaneous remedies. This undermines the research, but it's hard to fault patients trying to save their own lives.

Parents will have to contend with children who can find out about almost anything they want to, right from a home information appliance. Rating systems are already being designed to give parents control over what kids have access to, a topic I take up in chapter 12. This could become a major political issue if the information publishers don't handle it well.

On balance, I think the advantages of plentiful information will greatly outweigh the problems. More information means more choice. Today devoted fans plan their evenings around the broadcast times of their favorite television shows, but this will change once video-on-demand gives us the opportunity to watch whatever we like whenever we like. Family or social activities, rather than a broadcaster's time slots, will control our entertainment schedules. Before the telephone people thought of their immediate neighbors as their community. Almost everything they did was with people who lived close by. The telephone and the automobile allowed us to stretch out our communities. We may visit face-to-face less often than we did a century ago because we can pick up the telephone, but this doesn't mean we have become isolated. It's easier for us to talk to each other and stay in touch. Sometimes it may seem too easy for people to reach us.

A decade from now you may shake your head when you remember that there was ever a time when any stranger or a wrong number could interrupt you at home with a phone call. Cellular phones, pagers, and fax machines have already made it necessary for businesspeople to make explicit some decisions that used to be implicit. A decade ago you didn't have to decide whether you wanted to receive documents at home or take calls on the road. It was easy to withdraw to your house—or certainly to your car. With modern technology you have to decide when and where you want to be available. In the future, when you'll be able to work anywhere, reach anyone from anywhere, and be reached anywhere, you'll also have an easier time determining who and what can intrude. By explicitly indicating allowable interruptions, you'll be able to reestablish a sense of sanctuary.

Software on the interactive network will help by prescreening all incoming communications, whether live phone calls, multimedia documents, e-mail, advertisements, or even news flashes. Anyone you've approved will be able to get through to your electronic in-box or ring your phone. You might allow some people to send you mail but not to telephone. You might let others call when you've indicated that you're not busy and let still others get through anytime. You won't want to receive thousands of unsolicited advertisements every day, but if you send out a query for tickets to a sold-out concert, you'll want to get responses right away. You'll be able to have incoming communications tagged by source and type—for instance, answers to your queries, work-related documents, bills, greetings, inquiries, publications, ads. You'll set explicit delivery policies, deciding who can make your phone ring during dinner, who can reach you in your car or when you're on vacation, and which kinds of calls or messages are worth waking you for in the middle of the night. You'll be able to make as many distinctions as you need to and to change the criteria whenever you want. Instead of giving out your phone number, which can be passed around and used indefinitely, you'll add a caller's name to a constantly updated list that indicates your level of interest in receiving his messages. If someone not on any of your lists wants to reach you, he'll have to have someone who is on one of your lists forward the message. You'll always be able to move someone to a lower priority level or delete a name altogether from all lists. If you do that, the caller will have to send you one of those paid messages I talked about in chapter 8 if he wants to get your attention.

As you might deduce from the description of all this activity, your computer will rarely if ever be turned off. If it is, it will turn itself back on the instant a need arises. It will be a simple-to-use device at the center of entertainment, communications, and productivity in your office and at home. Your computer will interconnect with VCRs, stereos, TVs, security systems, and the Internet. You'll take it for granted as you use it to choose music, check a movie review, play a game, review your finances, or dig up a recipe. It will be as standard a piece of equipment in your home as the telephone.

These changes in home technology will start to influence residential architecture. Computer-controlled displays of various sizes will be built into the design of a house, and thought will be given to the placement of screens in relation to windows to minimize reflection and glare. Wires to connect components will be installed during construction. When information appliances are connected to the Internet, we'll have less need for reference books, stereo receivers, compact discs, fax machines, file drawers, and storage boxes for records and receipts. A lot of space-consuming clutter will collapse into digital information we can recall at will. We'll even be able to store old photographs digitally and call them up on a screen if we want to.

I've been giving these kinds of details a lot of thought because I'm building a house now, and I'm trying to anticipate the near future in its plans. My house is being designed and constructed so that it's a bit ahead of its time, and it might suggest things about the future of homes in general. I'll warn you, though—when I describe the plans, people sometimes give me a look that says, "You're sure you really want to do this?"

Like almost anybody who thinks about building a house, I want mine to be in harmony with the land it sits on and with the needs of the people who will live in it. I want it to be architecturally appealing, but I mostly want it to be comfortable. It's where my family and I will live. You can think of a house as an intimate companion, or you can look at the idea of a house through the eyes of the great twentieth century architect Le Corbusier, who said that a house is "a machine for living in."

My house is made of wood, glass, concrete, and stone. It's built into a hillside, and most of the glass faces west over Lake Washington to Seattle to take advantage of the sunset and views of the Olympic Mountains.

My house is also made of silicon and software. The installation of silicon microprocessors and memory chips, and the software that makes them useful, will let the house approximate some of the features that the interactive network will bring to millions of houses in a few years. The technology I'll use is experimental today, but over time some of it could become widely accepted and get less expensive. The house's entertainment system will be a close enough simulation of how media will work in the near future that I'll be able to get a sense of what it will be like to live with various technologies.

Of course, it won't be possible to simulate interactive broadband entertainment applications, because they require that a lot of other people be connected too. Having a private broadband network is a little like only one person having a telephone. The really interesting network applications will grow out of the participation of tens or hundreds of millions of people who will not just consume entertainment and other information but will create it too. Until a large share of the people in developed countries are communicating with each other on broadband networks, exploring subjects they have common interests in and making all sorts of multimedia contributions of their own, we won't feel the full effects of broadband interactive technology.

The cutting-edge technology in the house I'm building won't be just for previewing entertainment applications. It will also help my family meet the usual domestic needs for heat, light, comfort, convenience, pleasure, and security. It wasn't that long ago that the public would have been amazed at the idea of a house with electric lights, flush toilets, telephones, and air-conditioning. My goal is a house that offers entertainment and stimulates creativity in a relaxed, pleasant, welcoming atmosphere. These desires aren't very different from those of people who could afford adventurous houses in the past. I'm experimenting to find out what works best, and there's a long tradition of that kind of experimentation.

In 1925, when the newspaper magnate William Randolph Hearst moved into his California castle, San Simeon, he wanted the best in modern technology. Back then it was awkward and time-consuming to tune radio receivers to the various stations, so Hearst had several radios installed in the basement of San Simeon, each tuned to a different station. The speaker wires ran to Hearst's private suite on the third floor, where they were routed into a fifteenth century oak cabinet. At the push of a button, Hearst could listen to the station of his choice. Such ease of selection was a marvel in his day. Today it's a standard feature on every car radio.

I am certainly not comparing my house to San Simeon, one of the West Coast's monuments to excess. All I'm saying is that the technological innovations I have in mind for my house are not really different in spirit from those Hearst wanted in his. He wanted news and entertainment, all at a touch. So do I.

I began thinking about building a new house in the late 1980s. I wanted a house that would accommodate sophisticated, changing technology but in an unobtrusive way that made it clear technology was the servant, not the master. I didn't want the house to be defined by its use of technology. I was a bachelor when we did the original design, and when Melinda and I got married we changed the plans to make the house reflect her tastes and suit a family. Among other things, we redesigned the kitchen to accommodate a family better. The kitchen appliances, by the way, feature no more advanced technology than you'd find in any other well-appointed kitchen. Melinda also pointed out that I had a great study but there was no place designated for her to work, and the redesign took care of that deficiency too.

Computer rendering of the Gateses' future home, showing the view from the northwest across Lake Washington

I chose the property on the shore of Lake Washington partly because it was within easy commuting distance of Microsoft. In 1990 we began work on a guest cottage. Then in 1992 we began to excavate and lay the foundation for the main residence. This was a big job that called for a lot of concrete because Seattle is in an earthquake zone at least as perilous as California's.

Living space will be about average for a large house. The family living room will be about fourteen by twenty-eight feet, including an area for watching television or listening to music. The house will have cozy spaces for one or two people, although there will also be a reception hall to entertain one hundred for dinner comfortably—I enjoy having get-togethers for new Microsoft employees and summer interns. The house will also have a small movie theater, a pool, and a trampoline room. A sports court will be sited in the trees near the water, behind a dock for water-skiing, one of my favorite sports. A small estuary will be fed with groundwater from the hill behind the house. We'll seed the estuary with sea-run cutthroat trout, and we've been told to expect river otters.

Computer rendering of the Gateses' future home, showing the staircase and the formal dining room

If you come to visit, you'll drive down a winding driveway that approaches the house through an emergent forest of maple and alder punctuated by Douglas fir. Several years ago, decomposing duff from the forest floor of a logging area was spread across the back of the property. All kinds of interesting things are growing there now. After a few decades, as the forest matures, Douglas fir will dominate the site, just as it did before the area was logged for the first time at the turn of the twentieth century.

When you stop your car in the semicircular turnaround, you'll be at the front door, but you won't see much of the house. That's because you'll be entering onto the top floor. First thing, as you come in, you'll be presented with an electronic pin to clip to your clothes. The pin will connect you up to the electronic services of the house. You'll go down to the ground floor by either an elevator or a staircase that runs straight toward the water under a sloping glass ceiling supported by posts of Douglas fir. The house has lots of exposed horizontal beams and vertical supports. You'll have a great view of the lake as you go down to the first level, and I hope that the view and the Douglas fir, rather than the novelty of the electronic pin, will be what interest you most as you descend. Most of the wood came from an eighty-year-old Weyerhaeuser lumber mill out on the Columbia River that was being torn down. This wood was harvested nearly a hundred years ago and came from trees that were as much as 350 feet tall and between 8 and 15 feet in diameter. Douglas fir is one of the strongest woods in the world for its weight. Unfortunately, new-growth Douglas fir tends to split if you try to mill it into beams because the grain is not as tight in a seventy-year-old tree as it is in a five-hundred-year-old tree. Almost all of the old-growth Douglas fir has been harvested now, and any that remains should be preserved. I was lucky to find old-growth timbers I could reuse.

The fir beams support the two floors of private living space you'll go past on your way to the ground floor. Privacy is important. I want a house that will still feel private even when guests are enjoying other parts of it.

At the bottom of the stairs, the theater will be on the right, and to the left, on the south side, you'll see the big reception hall. As you step into the reception hall, on your right will be a series of sliding glass doors that open onto a terrace leading to the lake. Recessed into the east wall will be twenty-four video monitors stacked four high and six across, each with a 40-inch picture tube. These monitors will work cooperatively to display large images for art, entertainment, or business. I had hoped that when we weren't using the monitors they could literally disappear into the woodwork. I wanted the screens to display wood-grain patterns that matched the rest of the wall. Unfortunately, we could never achieve anything convincing with current technology because a monitor emits light while real wood reflects it. We settled for having the monitors disappear behind wood panels when we're not using them.

The electronic pin you wear will tell the house who and where you are, and the house will use this information to try to meet and even anticipate your needs—all as unobtrusively as possible. Someday, instead of your needing the pin, we might have a camera system with visual-recognition capabilities, but that's beyond current technology. When it's dark outside, the pin will cause a zone of light to move with you through the house. Unoccupied rooms will be unlit. As you walk down a hallway, you might not notice the lights ahead of you gradually coming up to full brightness and the lights behind you fading. Music will move with you too. It will seem to be everywhere, although in fact people in other parts of the house will be hearing entirely different music or nothing at all. A movie or the news will be able to follow you around the house too. If you get a phone call, only the handset nearest you will ring.

You won't be confronted by the technology, but it will be readily available to you. A handheld remote control will put you in charge of your immediate environment and of the house's entertainment system. The remote control will extend the capabilities of the pin. You'll use the remote to tell the monitors in a room to become visible and what to display. You'll be able to choose from among thousands of pictures, recordings, movies, and television programs, and you'll have all sorts of options available for selecting information.

A console, which will be the equivalent of a keyboard and let you give the house system very specific instructions, will be discreetly visible in each room, but it won't attract attention. A characteristic, easy-to-recognize feature will alert someone who walks into the room to the identity and whereabouts of the console. The telephone we all use has this quality already. It doesn't attract particular attention, but it's there when we need it.

Every computerized system should be so simple and natural to use that people don't give it a second thought. But simple is difficult. Computers do get easier to use every year, though, and trial-and-error in my house will help us learn how to create a system that's really simple to use. You won't have to make direct instructions and requests. You won't have to ask for a song by name, for example. You'll be able to ask the house system to play the latest hits, or songs by a particular artist, or songs that were performed at Woodstock, or music composed in eighteenth-century Vienna, or songs with the word "Yellow" in their titles. You'll be able to ask for songs you've categorized with a certain adjective or songs that haven't been played when a particular person was visiting the house before. You might program classical music as a background for contemplation and something more modern and energetic to play while you're exercising. If you want to watch the movie that won the 1957 Academy Award for best picture, you can ask for it that way—*The Bridge on the River Kwai* will come up on the screen. You could find the same movie by asking for films starring Alec Guinness or William Holden or films about prison camps.

If you were planning to visit Hong Kong soon, you might ask the screen in your room to show you pictures of the city. It will seem to you as if the photographs are displayed everywhere in the house, although the images will actually materialize on the walls of rooms just before you walk in and vanish after you leave. If you and I are watching different things and one of us walks into a room where the other is sitting, the house will follow mediating guidelines. The house might continue the audio and visual imagery for the person who was in the room first, or it might change programming to something it knows we both like.

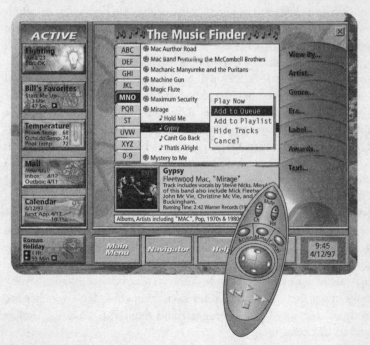

Prototype of a home control console

A house that tracks its occupants in order to meet their particular needs combines two traditions. The first is the tradition of unobtrusive service, and the other is that an object we carry entitles us to be treated in a certain way. You're already used to the idea that an object can authenticate you. It can inform people or machinery that you have permission to do something such as open a locked door, get on an airplane, or use a specific line of credit to buy something. Keys, electronic entry cards, driver's licenses, passports, name badges, credit cards, and tickets are all forms of authentication. If I give you the keys to my car, the car allows you to get in, start the engine, and drive away. You might say that the car trusts you because you carry its keys. If I give a parking attendant a key that fits my car's ignition but not its trunk, the car lets him drive it but not open the trunk. It's no different with my house, which will make various amenities available to you based on the electronic key you carry.

Nothing I'm planning is really so radical. Some visionaries foresee much bigger changes. They predict that within the next ten years robots will wander around our houses helping us out with various chores. I'm not preparing for that because I think it will be many decades before robots are practical. The only robots I expect to see in widespread use soon are intelligent toys. Kids will be able to program them in a limited number of ways to respond to specific situations and even to speak in the voices of favorite characters. They'll have limited "vision" and know the distance to the wall in each direction, the time, and the lighting conditions. They'll accept limited speech input, which will be a kick. (I think it would have been cool to have had a toy-size car when I was a kid that I could have talked to and programmed to respond to my instructions.) Other than toys, the other major robotic devices I see coming along soon will be for military applications. I doubt that intelligent robots will provide much help around the house in the foreseeable future. It takes a great deal of visual intelligence and dexterity to prepare food or change diapers. Pool cleaning, lawn mowing, and perhaps even vacuum cleaning can be done with a relatively dumb system, but once we get beyond tasks in which a robot just pushes something around, it's very hard to design a machine that can recognize and respond to all of the contingencies that come up.

The systems I'm building into the house are designed to make it easy to live in, but I won't know for sure if they're worthwhile until I move in. I'm experimenting and learning all the time. The design team used my guest cottage, which was built before the house, as a sort of test laboratory for home instrumentation. Because some people like the temperature warmer than others do, the cottage's software sets the temperature in reaction to who's inside and the time of day. The cottage knows to make the temperature toasty on a cold morning before a guest gets out of bed. In the evening the cottage's lights dim if a television is on. If someone is in the cottage during the day, the cottage matches its inside brightness to the brightness of the outdoors. Of course the occupant can always give explicit directions to overrule the settings.

This sort of instrumentation can be used to achieve significant energy savings. A number of electric utilities are testing a network to monitor the use of energy in individual households. If successful, this approach would end the expensive practice of having meter readers come to each household every month or two. But more important, computers in the home and at the utility company will be able to manage the minute-by-minute demand for power at various times of the day. Energy-demand management can save a lot of money and help the environment by reducing peak loads.

Not all of our experiments in the guest cottage were successful. I installed speakers that descended from the ceiling when they were needed. The speaker enclosures were suspended away from the walls, in an optimal acoustical position. But when I tried it out, it reminded me too much of James Bond gadgets, so in the main house we've settled for concealed speakers.

A house that tries to guess what you want has to be right often enough that you don't get annoyed by its miscalculations. I went to a party once at a house that had a computerized home-control system. The lights were set to go out at ten-thirty, which is when the owner usually went to bed. At ten-thirty the party was still going strong, but sure enough, the lights went out. The host was away for what seemed like a long time trying to get the lights back on. Some office buildings use motion detectors to control the lighting in each room. If there hasn't been any major activity for a few minutes, the lights go off. People who sit nearly motionless at their desks learn to wave their arms periodically.

-------------------------->

It isn't that hard to turn lights on and off yourself. Light switches are extremely reliable and very easy to use, so you run a risk when you start replacing them with computer-controlled devices. You have to install systems that work an incredibly high percentage of the time because your payoff in convenience can be eliminated by any lack of reliability or sensitivity. I'm hoping that the house system will be able to set the lights at the right levels automatically. But just in case, every room has wall switches we can use to override the computer's lighting decisions.

If you regularly ask for the light in the house to be unusually bright or dim, the system will assume that's how you want it most of the time. In fact, the house system will remember everything it learns about your preferences. If in the past you've asked to see paintings by Henri Matisse or photographs by Chris Johns of *National Geographic*, you may find other works of theirs displayed on the walls of rooms you enter. If you listened to Mozart horn concertos the last time you visited, you might find them on again when you come back. If you don't take telephone calls during dinner, the phone won't ring if the call is for you. We'll also be able to "tell" the house system what a guest likes. Paul Allen is a Jimi Hendrix fan, and a head-banging guitar lick will greet him whenever he visits.

The house will be instrumented so that it records statistics on the operations of all systems, and we'll be able to analyze that information to tune the systems. This presages an instrumented world.

As the Internet grows in popularity, instrumentation will count and keep track of all sorts of things, and the tallies will be published for anybody who's interested. We see precursors of this tabulation technology today. The Internet already carries information about local traffic patterns, which is great for deciding on alternative commuting routes. Television news programs often show traffic as seen by cameras in helicopters, and sensors embedded in the freeways estimate traffic speeds during rush hours.

We have student programmers on several college campuses to thank for trivial but amusing examples of such tracking instrumentation. At Carnegie-Mellon University, students have instrumented a soft-drink vending machine by connecting the hardware to the machine's empty-indicator light. The machine publishes information about its contents on the Internet continuously. It's a bit of frivolous engineering, but each week hundreds of people from all over the world check to see whether there's any 7UP or Diet Coke left in the vending machine.

The Internet already displays live video snapshots from many public places, lottery numbers and sports betting odds, current mortgage rates, and so on. Counts of crime reports and of campaign contributions by area and almost any other kind of public or potentially public information will soon be ours for the asking. I expect that eventually we'll be able to call up live pictures from various places around a city and ask for overlays to show spaces for rent with a list of the prices and the dates they're available.

I'll be the first home user of one of the most unusual electronic features in my house. The product is a database of more than a million still images, including photographs and reproductions of paintings. If you're a guest, you'll be able to call up on screens throughout the house almost any image you like—presidential portraits, reproductions of High Renaissance paintings, pictures of sunsets, airplanes, skiers in the Andes, a rare French stamp, the Beatles in 1965.

A few years ago I started a small company, now called Corbis, to build a unique, comprehensive digital archive. Corbis is a digital stock agency for images of all kinds. In late 1995 Corbis expanded its stock of images when we bought the Bettmann Archive, a leading photo collection. Using high-quality scanners, Corbis converts images into digital form. The images are stored at high resolution in a database that has been indexed in inventive ways to make it easy for clients to find exactly the images they need. The digital images are available to commercial users such as magazine and book publishers and to individual browsers. Corbis has produced some multimedia products based on its archives too. Royalties are paid to the image owners, whose rights Corbis vigorously protects. Corbis is working with museums and libraries as well as with a large number of individual photographers, agencies, and other archives. In the spring of 1996, for instance, it acquired exclusive digital rights to the work of famed black-and-white photographer Ansel Adams.

I believe that quality images will be in great demand on the interactive network. My idea that the public will find image-browsing a worthwhile pursuit remains to be demonstrated, but I think that with the right interface the service will appeal to a lot of people. I'm looking forward to being able to ask for "sailboats" or "volcanoes" or "famous scientists" and then seeing what turns up.

Although some of the images will be of artworks, that doesn't mean I believe that reproductions are as good as the originals. In the course of my business travels, I've been able to spend time in museums looking at great art, and there's nothing like seeing the real work. The most interesting piece of "real art" I own is a scientific notebook kept by Leonardo da Vinci in the early 1500s. I've admired Leonardo ever since I was young because he was a genius in so many fields and so far ahead of his time. Even though it's a notebook of writings and drawings rather than a painting, no reproduction could do this work of art full justice either. Still, I believe that easy-to-browse image databases will get more people interested in both graphic and photographic art.

Art, like most things, is more enjoyable when you know something about it. You can walk for hours through the Louvre admiring paintings, but the experience becomes much more interesting when someone knowledgeable is walking with you. A multimedia document can play the role of guide whether you're at home or in a museum. It can let you hear part of a preeminent scholar's lecture on a work. It can refer you to other works by the same artist or from the same period. You can even zoom in for a closer look. If multimedia reproductions and presentations make art more accessible and approachable, people who see the reproductions will want to see the originals. Exposure to the reproductions is likely to increase rather than diminish reverence for the real art and encourage more people to get out to museums and galleries.

A decade from now access to the millions of images and all the other entertainment opportunities I've described will be available in many homes and will certainly be more impressive than those I'll have when I move into my house. My house will just be getting some of the services a little sooner.

I enjoy experimenting, and I know that some of my concepts for the house will work out better than others. Maybe I'll decide to conceal the monitors behind conventional wall art or throw the electronic pins into the trash. Or maybe I'll get fond of the systems in the house and wonder how I ever got along without them. That's my hope.

11

THE INTERNET
GOLD RUSH

At the height of information highway mania, in 1994 and much of 1995, it seemed as though almost every day some company or consortium staked a claim in the race to build broadband interactive networks that would deliver video entertainment into residential neighborhoods. Incessant hoopla over megamergers and other bold investments created a gold rush atmosphere—people and companies rushed headlong to lay claim to technology and content they hoped would pay off.

Investors were enchanted with highway-related stock offerings. Media coverage was unprecedented, especially when you consider that both the broadband technology and consumer demand for interactive entertainment services were unproven. The urgent atmosphere of almost panicky optimism made the early, unchronicled days of the personal computer industry seem sedate by comparison. The frenzy was intoxicating.

Many of the investment and merger announcements seemed to be motivated by companies' wanting to stay ahead of other companies rather than by an understanding of the economics. In retrospect it is clear that the gold lay deeper and in different places than the would-be prospectors realized. Most telephone and cable television companies had been fixated on video-on-demand, and some were intrigued by the idea of video telephones. They didn't foresee much of a role for the personal computer.

It was difficult to guess reliably what kinds of interactive applications would appeal to the public, although I knew how I would use a broadband network to pursue my interests. I like to stay up-to-date on medical advances, for instance. I like to find out how to minimize health risks for someone in my age group. So I'd want a broadband network to deliver fitness and medical information as well as applications that would help me educate myself in other areas. But that was just me. Would other users want medical advice? New kinds of games? New ways of meeting people? Home shopping? Or just a lot more movies? No one knew.

Faced with this uncertainty, companies in the telephone, cable television, and software industries planned dozens of field trials of broadband services in North America, Europe, and Asia. The companies were willing to underwrite these expensive exercises to get a head start in validating software designs and figuring out which applications and services would appeal to consumers. Field trials were the only way to build a case for the huge investment required to bring interactive television to residential communities. The business models were uncertain, and making a wrong decision could waste huge sums of money. A surprising number of companies committed to moving at full speed while others proceeded more cautiously.

Then the Internet blossomed into public awareness as it achieved critical mass.

Almost overnight the Interactive Television Gold Rush all but died, and the Internet Gold Rush was born. Suddenly interactive TV was passé, and interactive, networked computing was hot. It was a big change for companies to realize that consumer involvement with interactive services would take place first on the PC and only later involve the TV and other devices. The swiftness of the change was breathtaking. It caught me off guard—even though the increasing popularity of the PC was critical to the shift. But it was a welcome development. It fit Microsoft's original vision of "a personal computer on every desk and in every home." The Internet will deliver on our "information at your fingertips" vision too.

Strategies changed abruptly at every communications, computer, software, and content company that had been staking claims along the expected route of the information highway. Many plans were summarily abandoned as the spiraling success of the World Wide Web demonstrated that people would pay for connectivity to interactive content of many kinds—even if that content was delivered on maddeningly slow narrowband connections. Almost all of the broadband trials were canceled. They were much less important now that millions of people were already demonstrating their enthusiasm for interactivity—and declaring their specific interests—by logging countless hours on Web sites. This vibrant near-term market for interactive connectivity meant that interactive killer applications could evolve alongside increases in bandwidth to the home. Hundreds of new companies became Internet service providers, thousands of new corporate Web sites appeared each week, and computer hardware and software companies began creating Internet products.

The industry had assumed that residential network services would begin with entertainment applications such as movies-on-demand and then evolve through shopping to a fuzzy category of "interactive information services." The Internet stood this progression on its head. Because the Internet couldn't easily or securely handle the exchange of money, electronic commerce was delayed and most information on the net ended up free. As people were drawn by the millions to this incredibly diverse resource, "interactive information services" became the first rather than the last broad category of killer applications. As Internet software improved, commerce became feasible, and it promises to be an important Internet activity. Eventually, as true broadband networks reach homes, we'll see the entertainment applications that were the basis for the early optimism about interactive television.

People in a gold rush often get so caught up in the dream of quick riches that they overinvest in the obvious areas and ignore subtler or longer-term opportunities. In a real gold rush, the fever almost invariably leads to financial catastrophe for many of the prospectors—although some win big and the overall economy tends to benefit.

Few fortunes were made in the California gold fields a century and a half ago. There wasn't nearly enough gold to go around, and massive inflation made life particularly hard for prospectors who failed to cash in. A boiled egg cost 75 cents, many times what it did before the gold rush started in 1849. Most of the winners weren't miners, and most of the successes didn't happen overnight. A German immigrant named Levi Strauss set up a dry goods business in San Francisco three years after the gold rush started. Twenty years later, in 1872, a Nevada tailor offered Strauss a half-interest in a patent for riveted denim pants. Since then more than a billion Levi's jeans have been sold—blue gold, we might say.

The lasting legacy of the California Gold Rush was the economic development of the west by pioneers who had an optimistic can-do attitude that shaped the American psyche. Gold fever was the bootstrap that pulled California's economy up, enabling it to evolve rapidly. In 1848 California had an agricultural economic base that attracted only 400 settlers. The next year, with the onset of gold fever, 25,000 settlers poured in. A mere decade later manufacturing was a much bigger part of California's economy than gold production, and the state's per capita wealth was the country's highest.

The Internet Gold Rush is a force for economic progress. It's generating an unnaturally high level of investment as companies try out different approaches. Every major limitation of the Internet, from its questionable security to its chaotic content, has spawned dozens of companies working to solve the problem. Lots of money will be lost along the way. What seem to be lucrative niches today may wind up as highly competitive markets with low margins. Or they may simply turn out not to be popular. It's remarkable how high the valuations of Internet-related companies are, considering how difficult it is to achieve differentiation and profitability. The total revenue from Internet activities has been small, and the profit has been a large negative number, but people and companies keep at it because they have faith—or the fever.

A fair number of implausible promises have developed around the Internet. One radio ad I heard implies that you too can make a fortune from the Internet *even if you've never used a computer!* Just pay for a seminar, and the secret is yours. In press releases and news accounts, more sophisticated but equally improbable statements seem to promise remarkable benefits from this or that computer hardware or software product. I'm surprised by how uncritical people with Internet fever can be about some of these promises. If a company had announced back in 1994 that it would soon release a software program that was many times better than anything already on the market, most people would have been skeptical—and rightly so. But two years later somebody making the same announcement, but with the embellishment ". . . for the Internet," would encounter a surprising open-mindedness: "Software that's ten times better than anything out there now, and for the Internet? Wow!"

It pays to be wary of hype. Investment success is no more certain now than it was in the early days of the PC industry, when dozens of companies offered distinctly different kinds of PCs. If you had invested in Compaq, you'd be quite happy now. If you had invested in Mindset or Gavelin or Eagle Computer, each a high-flier in its day, you'd be unhappy. These companies probably seemed like good investments—they were building PCs for the PC revolution—but the competition was too stiff for most entrants. As is often the case, some of the best investments turned out to be in companies that pursued less obvious opportunities. Intel provided microprocessors, Microsoft supplied software, and Hewlett-Packard manufactured printers. Makers of other PC components, such as memory chips, have done well for investors too, although sometimes only intermittently because component markets are so volatile. Many companies have done poorly despite high hopes, capital, and talent. History will repeat itself. When the Internet frenzy is behind us, as it will be sooner or later, we'll be impressed by the huge successes and incredulous at the wreckage of failed ventures. "Who funded those companies?" we'll wonder. "What was going on in their minds? Was that just mania at work?"

It should be clear by now that, although I'm cautious about most Internet investments, I believe in the long-term opportunities afforded by interactive networks. Virtually everything Microsoft does these days reflects my conviction that the Internet is going to grow so that almost everyone in the developed world and huge numbers of people in the developing world will be users. Companies in most fields ignore it at their peril.

There are many distinct areas of Internet investment. Cable, telephone, and other companies will compete to provide the fiber, wireless, and satellite infrastructure. Hardware companies will compete to sell servers, ATM switches, set-top boxes, PCs, digital TVs, videophones, and other information appliances to consumers. A relatively small number of software companies will compete to deliver components of the Internet software platform, while a much larger number—many thousands—will produce Internet content, some of which will create new markets. Eventually millions of companies and individuals will sell information and entertainment across the net.

Cable television and telephone companies around the world will move along four parallel paths. First, each will be going after the other's business. Cable companies will offer telephone service, and phone companies will offer broadcast video. Second, both kinds of systems will provide midband connections to the Internet using ISDN, ADSL, or cable modems (all of which are described in chapter 5). Third, both cable and telephone companies will convert to digital technology in order to provide more television channels and higher-quality signals. Fourth, a few companies will conduct trials of broadband systems connected to television sets and PCs in affluent cities. Each of the four strategies will motivate investment in digital network capacity, spurring intense competition between phone companies and cable companies to be the first network provider with abundant bandwidth in a neighborhood.

Until 1996 a broadband interactive network that carried both video and telephone services was illegal in the United States. Phone companies couldn't transmit video, and cable television companies couldn't offer telephone services without submitting to overburdensome regulations. By the summer of 1995, though, Congress was embroiled in a debate about how the U.S. telecommunications industry should be deregulated. Early in 1996 Congress passed the Telecommunications Reform Act, freeing telephone, cable, and other companies both to compete with each other and to build the midband and ultimately broadband networks that will carry entertainment as well as communications.

In the immediate aftermath of deregulation, we saw lots of maneuvering among the telephone (local and long-distance), cable, and entertainment companies—complaints, lawsuits, blocking actions, and consolidations. Some companies were trying to slow down competitors getting into their businesses and at the same time speed their own entries into other companies' businesses.

Local phone companies are in a defensive position. They will face increased competition from phone and cable companies that will move in to offer local telephone and other communications services. Deregulation is unleashing this competition just a few years before the cost of long-distance voice-telephone service may drop as a result of competition from the Internet.

By 1997 communications companies will be installing broadband connections into all medium and large businesses, and midband connections into homes and small businesses. The opportunity to offer ISDN to residential PC users will provide new revenues to phone companies that bring prices down to establish a mass market. ADSL will become a big factor too, replacing ISDN in some cases. Using ADSL, phone companies will deliver telephone, video phone, and Internet connectivity into homes at good speeds. But ADSL isn't a full solution for the delivery of high-quality video because it doesn't offer enough bandwidth to feed different programming to multiple TV sets in a house—and in some places won't be able to offer enough bandwidth to feed even a single TV set a high-quality signal.

The phone companies are stronger financially than cable companies even though their shareholders expect large dividends. The U.S. local telephone exchange market alone, with annual revenues of about $100 billion, is far more profitable than the $20 billion U.S. cable business. The regional Bell operating companies (RBOCs) will compete with their former parent, AT&T, to provide long-distance, cellular, and new services. But like other phone companies around the world, which share their heritage as regulated utilities, the RBOCs are new to the competitive world.

The cable companies are younger and smaller than the big telephone companies and tend to be more entrepreneurial. Cable television networks provide customers with one-way broadband video through a web of coaxial, and sometimes fiber-optic, cable. Although their penetration worldwide is quite low—189 million subscribers—cable systems run past nearly 95 percent of all U.S. homes and into 68 million of them. Cable companies will go as fast as they can to upgrade their plants to support Internet connectivity at midband speeds. Only about 10 percent of the plant is ready today, and it will probably be five years before 75 percent is upgraded to handle midband traffic.

Worldwide, the cable and telephone industries will be the primary but not the only competitors to provide network connections. Railroad companies in Japan, for example, recognize that their rights-of-way for tracks would be ideal for long fiber-optic cable runs. Electric, gas, and water utilities in many countries point out that they too run lines into homes and businesses. Some of them have argued that the economies to be realized from computerized management of home heating, for instance, could defray much of the cost of stringing fiber-optic cables to the same homes. Centralized management, in which the utility's computer would influence or even control the energy use in thousands of homes, would level peak energy demands and thereby reduce the need for expensive new generating plants. As it happens, most of the cable TV connections in France are owned by two big water companies. But outside France, water and the other utility companies seem to be less obvious candidates for building communications networks.

The infrastructure will develop at different paces in different communities and in different countries. Networks will tend to go into richer neighborhoods first because that's where residents are likely to spend more. Local regulators may find themselves competing to create favorable environments for early deployment of the network, but no direct investment of taxpayer money will be needed to build the network infrastructure in industrialized countries that have procompetition regulations.

When I travel abroad, the foreign press often asks how many years behind developments in the United States their country is. It's a difficult question. Advantages in the United States are the size of the market, the popularity of the PC in American homes, and the competition between the phone and cable companies. U.S.-based companies are leaders in almost every technology that will be a part of building the broadband infrastructure: microprocessors, software, entertainment, personal computers, set-top boxes, and network-switching equipment. The only significant exceptions are display technology and memory chips.

Although the interest in high-tech communications systems is probably greater in Japan than in any other country, it's difficult to predict how much a leader Japan will be in using interactive networks. The use of personal computers in businesses, schools, and homes in numbers comparable to those for PC use in other developed countries was a relatively late development in Japan, although beginning in 1995 the market grew dramatically. The country was slow to get started with PCs partly because of the difficulty of entering kanji characters on a keyboard but also because of Japan's large, entrenched market for dedicated word processing machines.

Japan is second only to the United States in the number of its companies investing in developing highway infrastructure building blocks. Many large Japanese companies have excellent technology and a history of taking long-term approaches to their investments. NEC's corporate slogan, "Computers & Communication," anticipated the communications revolution as early as 1984 and is an indication of the company's commitment. Several Japanese companies combine technology expertise with content business. Sony owns Sony Music and Sony Pictures, which includes Columbia Records and Columbia Studios. Toshiba has a large investment in Time Warner. The cable industry in Japan was overregulated until quite recently, but the rate of deregulation has been impressive. The Japanese phone company, NTT, has one of the largest valuations among public companies in the world and will play a leadership role in developing the interactive network in Japan.

In Canada, as in the U.S., a high percentage of homes have cable television. This is an advantage because competition between cable companies and phone companies will accelerate the pace of investment in network infrastructure. Great Britain, however, is the furthest along in actually using a single network to provide both telephone and cable services. Cable companies there were allowed to offer phone service beginning in 1990. Foreign companies, primarily U.S. phone and cable companies, followed up with major investments in fiber infrastructure in the United Kingdom. Now British consumers can choose to get telephone service from their cable TV company. This competition has forced British Telecom to improve its rates and services.

In France, the pioneering on-line service, Minitel, has fostered a community of information publishers and stimulated the population's familiarity with on-line systems generally. Even though both terminal capability and bandwidth are limited, Minitel's success has encouraged innovation and provided lessons on the market.

In Germany, Deutsche Telekom lowered the price of ISDN service dramatically in 1995. This led to a significant increase in the number of users connecting up personal computers. Bringing ISDN prices down was clever because lower prices will promote the development of applications that will help hasten the arrival of a broadband system.

The level of PC penetration in business is even higher in the Nordic countries than in the United States. These countries understand that their highly educated workforces will benefit from having high-speed connections to the rest of the world.

On a visit to Sweden in 1996, I was impressed by Stockholm's aggressive deployment of a fiber network that leases capacity to private businesses including cable companies. Perhaps even more impressive was the fervor with which the city was building interactive applications, including the creation of an electronic marketplace for services. Mayor Mats Hulth issued a challenge to other large European cities to see which could use information technology most successfully. By the summer of 1996 twenty-three cities were competing in the "Bangemann Challenge," fielding ninety-nine projects in areas such as telecommuting, distance learning, university networks, road traffic control, and health care. The cities undertaking projects were Amsterdam, Antwerp, Barcelona, Berlin, Bologna, Bradford, Bremen, Brussels, Budapest, Edinburgh, Eindhoven, Gothenburg, The Hague, Hanover, Helsinki, London-Lewisham, Lyon, Manchester, Modena, Paris, Rotterdam, Stockholm, and Utrecht.

Eastern Europe isn't ignoring the communications revolution either. Modern telecommunications equipment has been installed in what was once East Germany. When I go to places such as Poland, the Czech Republic, and Hungary, I meet with telephone company and government officials, encouraging them to adopt policies that favor telecommunications investment. I find receptive audiences because many of these people are quite savvy about how global communications systems will help their countries create good jobs for their talented people.

Because of their great geographic distance from other developed countries, Australia and New Zealand are particularly motivated to pursue the opportunities high-speed networks offer. Australia's government-run PTT—a postal, telephone, and telegraph services monopoly—has been replaced by full open competition. This development encouraged forward-looking plans for the rapid buildout of infrastructure based on fiber and cable. The competitive environment is so fierce that many Australian homes and businesses may end up being served by competing connections to high-speed networks. New Zealand has the most open telecommunications market in the world, and its newly privatized phone company has set an example of how effective privatization can be. New Zealand went all out to deregulate computing and telecommunications. In the 1980s it reduced its high tariff on imported computers and privatized New Zealand Telecom in a successful effort to encourage foreign investment and competition. The U.S. telephone companies that own New Zealand Telecom use New Zealand as a test market for equipment and services.

In Singapore, the population density and the government's focus on infrastructure make it certain that this nation will be a leader in developing infrastructure. A decision by the Singaporean government to make something happen means quite a lot in this unique city-state. Network infrastructure is already under construction, and Singapore Telecommunications Ltd. is working with Time Warner on an eighteen-month trial of video-on-demand services. Residential developers in Singapore will be required to provide every new house or apartment with a broadband cable in the same way they are required by law to provide lines for water, gas, electricity, and telephone. When I visited with Lee Kuan Yew, the senior minister who was the political head of Singapore from 1959 to 1990, I was impressed by his understanding of the opportunity and his belief that it is a top priority to move ahead at full speed. Mr. Lee said he views it as imperative that his small country continue to be a premier location in Asia for high-value jobs. I was quite blunt in asking Mr. Lee if he understood that the Singaporean government would be giving up the tight control over information that it exercises today as a way of ensuring shared values that tend to keep societal problems in check. He said yes, that Singapore recognizes that in the future it will have to rely more on methods other than censorship to maintain a culture that sacrifices some Western-style freedom in exchange for a strong sense of community. It will be interesting to see how Singapore follows through on this.

Hong Kong's government provides infrastructure, seed money, and even rent discounts and other office support for private companies

----------------------------▶

innovating in electronics, telecommunications, and multimedia. The government seems committed to building a broadband network connecting the entire city, and it is allowing competition intended to encourage the development of services and content. The competition from private companies that Hong Kong Telecom will face for the first time is expected to result in the introduction of advanced services, such as interactive multimedia, that tie-in with Hong Kong's film and toy industries. Of course, no one is quite sure what will happen in Hong Kong after China takes control in July of 1997.

In China itself, the government seems to have a love-hate relationship with the potential of communications technology to serve the country. Authorities are trying to filter the electronic flow of information into China. Post and Telecommunications Minister Wu Jichuan told reporters at a news briefing, "By linking with the Internet we don't mean absolute freedom of information. I think there is general understanding about this. If you go through customs, you have to show your passport. It's the same with management of information." China's solution may turn out to be development of what amounts to a national intranet, somewhat isolated from the rest of the world's electronic communications.

Although the speed with which the infrastructure is brought directly into homes will correlate in large part with the per capita gross domestic product (GDP) of a country, even developing countries will start to see connections into businesses and schools that will have a huge impact and start to reduce the income and technology gaps between societies. Caribbean countries such as Jamaica and the Dominican Republic are already linked to North American countries by fiber cable so that their low-cost data entry services can be on-line. In Latin America, Costa Rica stands out as both a provider of data entry services and a host to multinational corporations. Bangalore in India, and Shanghai and Guangzhou in China will install network connections to businesses and universities that will offer the services of their highly educated workers to the global market.

Throughout the world, ground-based and satellite-based wireless services have important roles to play. Many information appliances will connect directly or indirectly to the network via infrared or radio links, enabling mobile access to narrowband signals. Current satellites offer high-quality broadcast video although there isn't enough bandwidth to provide customized streams of video at economical rates, and there isn't a good back channel. A partial solution is to use the telephone system for a narrowband back channel. Direct-broadcast satellite systems such as the Hughes Electronics DIRECTV system use your regular home telephone line to keep track of any pay-per-view programs you have chosen. DIRECTV is primarily an entertainment system—a wireless precursor of the "500 channel" future of cable television. But with a special add-in circuit, direct-broadcast satellites can send data to both PCs and television sets.

Teledesic, a company that cellular telephone pioneer Craig McCaw and I have invested in, is working on an ambitious plan to use a large number of low-orbit satellites to provide two-way broadband service to the whole world. Global communications coverage is achieved today by putting a handful of extremely expensive satellites in geosynchronous orbits 22,300 miles above Earth. This high orbit works for satellites that broadcast data, including television programming. But using the high orbit also requires a lot of power and entails a significant transmission delay, factors that make two-way communications through a geosynchronous satellite problematic.

Teledesic proposes to loft several hundred relatively inexpensive satellites into polar orbits a mere 435 miles up—fifty times closer to Earth than the geosynchronous satellites. This relative proximity to Earth means that the Teledesic satellites would need 2,500 times less power than high-orbit satellites, that they would have reliable two-way capability, and that they would be able to provide transmission speeds comparable to those available with fiber. It's an exciting idea.

Teledesic has clearance for the frequencies it needs to operate in most countries, but it still faces major technical and financial challenges. Part of what makes the Teledesic vision so bold is that the company can't scale its plans back to a small number of satellites. Unlike satellites in geosynchronous orbit, Teledesic's satellites would move relative to the surface of the Earth. To provide continuous service to even a single geographic area would require a lot of satellites to be in orbit so that at least one was passing overhead at all times. With enough satellites aloft, the

Teledesic system could provide global communications coverage—the same quality and quantity of interactive broadband service available everywhere.

Teledesic may discover that it makes most of its money serving the rural parts of developed countries. There isn't likely to be a lot of revenue from the less developed countries or from satellites passing over the oceans. I can imagine Teledesic, or any other company deploying low-orbit satellites, entering into creative flat-fee deals that let the governments of developing countries use the bandwidth as they see fit. Service at sea might be sold inexpensively too, since there won't be much demand there for the Teledesic system's abundant bandwidth. Once the system is fully deployed, its communications capacity will be as great over central China or the central Pacific as it will be over upstate New York, and the costs of running the system will be the same whether it's lightly or heavily used. These economics resemble the commercial realities an airline faces: A passenger plane costs almost as much to fly with its seats empty as full, so why not sell otherwise unused capacity at bargain rates?

If it succeeds, Teledesic could transform millions of lives by bringing state-of-the-art communication and the Internet's educational, commercial, and entertainment content to an audience that might not be able to connect up otherwise. It's easy for those of us in developed countries to forget that two-thirds of the world's people have never even made a phone call. Where telephone service is available, it's provided over analog, copper-wire networks that are often antiquated and unlikely to be upgraded to a digital, broadband capability any time soon outside the most developed markets. While many places in the world are connected by fiber, so far it is used primarily for the trunk lines between countries and for telephone company switches. Fiber to homes and offices is expensive to install even in high-density areas, and it is unclear when it will ever be feasible to justify running fiber extensively in low-density areas.

Even without Teledesic or other proposed satellite systems, the Internet's audience will grow into the hundreds of millions. That's one reason the software industry has an overwhelming focus on the Internet now.

The highest profile software competition as we head into 1997 is in "Web browsers," programs that enable you to browse pages on the World Wide Web or corporate intranets. The most notable browsers are Netscape Navigator, which got a large early following because it led the way with browsing features and was at one time free, and Microsoft Internet Explorer, which remains free.

Netscape has been the leader in this new category of software, and Microsoft's strategy for catching up and moving ahead is to "embrace and extend." This is the same approach Lotus used with Lotus 1-2-3 in the early 1980s to displace the pioneering spreadsheet VisiCalc, which had been the commanding market leader. It's the strategy Microsoft in turn used with Microsoft Excel to displace 1-2-3. Each new version of Microsoft Internet Explorer embraces all of the popular Internet browsing features, including features developed by Netscape, and then extends functionality with new features aimed at both Windows and Macintosh users. Microsoft calls its Internet technologies "ActiveX," a name that reflects the belief that static, lifeless Web pages are boring and that popular Web sites will come "alive" with interactivity. Of course, Netscape also embraces and extends whenever it releases a new version of Navigator. Extensions of these kinds, from whichever party, become the standards that the other company in turn embraces—as long as users and publishers find them attractive.

The role of Internet publishers, or "Webmasters," is important since they determine which of the new extensions get used to create great Web sites. Netscape Navigator version 2.0 introduced support for frames, independent regions of the screen that have rectangular borders and can display Web pages within Web pages. Some Webmasters decided that frames were valuable and used them on their sites. Microsoft Internet Explorer 2.0 didn't have this capability, so people browsing a Web site that took advantage of frames had an incentive to use Navigator instead. Netscape was ahead at that point. But Microsoft came back with Internet Explorer 3.0, which among other things embraced the frame protocols and extended them to allow borderless, custom-bordered, and floating frames. Meanwhile Netscape was hard at work on its own enhancements.

The existence of a variety of browsers presents a challenge to Web site publishers, just as having a variety of operating systems presents a challenge to software developers: Which standards should a publisher support? Besides all the different browsers and all of their versions, publishers have to consider hardware limitations such as screen size, memory size, audio capability, and graphics capabilities. The rate at which Web site publishers will be willing to take advantage of new protocol extensions is limited by their budgets. There can't be twenty new extensions every year forever. A fair number of new protocols are still needed, however, because Web pages circa 1996 are not rich enough to suit publishers' and ultimately consumers' expectations.

A consequence of this innovation is that browsers are growing into large programs that engage every element of a PC. Now that browsers support Web pages that use Java programming, Shockwave animations, 3D imagery, Acrobat files, compressed graphics, video, audio, rich fonts, and so on, they have become more demanding of a computer's resources, including memory and processing power, than any other popular program.

The competition between operating systems will be greatly influenced by how well each integrates browser features. Microsoft's strategy is to build Web access right into Windows. Netscape's strategy is to make Windows and the Apple Macintosh operating system irrelevant by developing Navigator into a full-featured operating system.

No matter who wins the contest, the result will be a significant advance: combining the best of the PC with the best of the Web, creating a single way of dealing with information. Local and remote data will be handled identically. Browsing active pages by means of links will be the interface you use—regardless of whether you're looking at data stored on a CD-ROM, a hard disk, or a faraway Internet site.

Windows 95 allows a folder (called a "directory" in older versions of Windows) to contain links to files and other folders. These links are called "shortcuts." Double-clicking on one takes you to wherever the shortcut leads. These shortcuts are just like links on Web pages. The next step is to allow anything used on Web pages to be used in folders so that folders can provide more information than just lists of the filenames they contain. Folders will include descriptions of files, images, animations, borderless frames, and any other Web feature. "Web view" is the name

A folder displayed in the Windows Web view, in which a single click takes you to any file or folder

for these rich folder views. Web view is a feature of Internet Explorer 4, which will come out by early 1997.

As part of its strategy to make Navigator an operating systems platform, Netscape will enhance Navigator with new interfaces for memory management, file systems, security, scheduling, graphics, and everything else that applications require in an operating system such as Windows. The plan is to make Netscape's browser a de facto software platform that sits on top of Windows, the Macintosh operating system, or UNIX. Netscape hopes people will discard their existing applications in favor of new ones that will evolve around Netscape's standards.

This general approach to competing against an operating system has been tried before. Lotus had aspirations for its Notes software to become an important middleware platform on which mainstream applications would be built. But the prospects for Notes as middleware have been diminished by the rise of the Internet. It's tough to achieve a huge success with a middleware strategy when the most popular operating systems are embracing, in an integrated way, the same features that the middleware promotes.

Another competition engendered by the popularity of the Internet is determining how PCs and terminals should change to be cheaper and more suitable for browsing. Some observers and interested parties believe that the important face-off is between the PC itself and a combination of servers and limited-purpose terminals—the servers being more expensive than PCs, and the terminals less powerful and less expensive than PCs. The main proponents of this "strong-server/weak-client" model are Sun Microsystems, a company that sells servers, and Oracle, a company that sells server software. Sun and Oracle contend that the Internet will finally do for them what they've failed to do in the past—reverse the movement toward powerful personal machines and recentralize computing.

Sun and Oracle call the client machines they champion "Internet terminals," or "network computers." These devices don't have disk drives and instead will draw their software and information from centralized servers. The companies say that the combination of servers, downloadable software, and fast communications links will make real PCs unnecessary for many people. They maintain that these networked terminals, which are by design incompatible with today's PCs and applications, are all that most people need.

As I write this in the summer of 1996, all that is clear is that there will be a lot of different devices incompatible with PC software. For each device, it's important to consider what is left out and what the trade-offs are.

If there's one thing PC users have shown us, it's that people don't want to settle for outdated features or performance. The consumer's appetite for an improved experience is what drives innovation and makes three-year-old PCs about as popular as three-year-old newspapers. The continuing aggressive rate of improvement in the Internet all but guarantees that diskless terminals will be disappointing once they are even a few months old—unless they're connected to very fast networks.

Many of my reservations about diskless terminals go away if the network to which they're connected is fast enough. Even large software programs can be downloaded in seconds if the available bandwidth from the network server is great enough. But not all diskless PCs will be equally useful. The best choices will be the ones that give users lots of control and that run popular applications.

Sun and Oracle used to promote diskless "X-terminals" for corporate local-area networks. These so-called "dumb terminals" never caught on, in part because there wasn't much of a savings compared to the cost of a real PC.

The future lies in balanced client/server computing in which both the client (usually a PC) and the server are quite capable and cooperate in running software applications. I don't see much reason to dumb down the desktop machine. And I don't see the justification for businesses to invest heavily in new servers based on mainframes or clusters of mini-computers. Technology is making PC servers with great performance and low prices possible.

The supposed advantages of dumb terminals will be equally available to people who invest a little more and get a personal computer that has a disk drive and can run any of the thousands of Windows-based applications. Entry-level PCs are now about $1,000 or $1,200 in the United States, but the price could approach $600 as manufacturers direct some of their innovative energy into lowering prices for the mass market rather than just increasing performance. PCs and lots of other information appliances, including TV set-top boxes and hand-held devices, will connect to the Internet and sell in large volumes. But I'm unenthusiastic about the prospects of incompatible dumb terminals that are almost, but not quite, real PCs.

Supplying Internet connections, browser software, operating systems, servers, and information appliances are opportunities for technology companies. But supplying Internet content is a potential opportunity for every company. No company is too small to participate. Although the television revolution that began fifty years ago generated some profits for technology companies manufacturing TV sets, the biggest winners were the companies that used the new medium to deliver content: entertainment, news, sports, and advertising. We'll see a similar dynamic for the Internet. This time around anybody can be an information publisher because anybody can get a "channel" for their content—a Web site. We'll see intense competition in all categories of popular content, including news, commentary, sports, games, directories, and classified advertising.

Investments in Internet publishing to date have not generated much income or profits. The biggest payback so far has come from using the Internet to exchange information, including marketing information. Optimism that profits will appear someday is widespread, although I think it will take a lot longer than most investors seem to believe.

The most promising source of eventual revenue from the Internet is advertising, a proven money-maker in conventional media. Advertisers tend to be reluctant to enter a new medium, but people at many advertising agencies are enthusiastic about pioneering ways to exploit interactivity. Narrowband Internet users get annoyed at having to wait for big advertising images to download, but as midband connections become commonplace, downloads will be far faster and advertising won't carry big performance penalties.

Some content companies are trying to charge subscription fees for browsing their Web content. It's a tough sell, though, because there are immense amounts of free content on the Internet. Sites that offer no free content get relatively few visitors, which makes it hard to attract advertising.

Another funding source for content creators is the on-line services, including America Online, CompuServe, Prodigy, and MSN, who will pay content entrepreneurs to deliver content exclusively to their subscribers.

A variety of Internet business models will be tried, and some will prove themselves. Because content is an attractive business in other media, there will be a huge number of Internet content entrants, large and small. Just as Microsoft is investing in content, many other companies are too. The ambitions of most cable and phone companies, for example, go well beyond simply providing a pipe for delivering bits. Many communications companies want a financial interest in some of the bits being shipped. They see the network-based economy as a kind of food chain, with the delivery and distribution of bits at the bottom and various types of applications, services, and content layered on top. Companies in the bit-distribution business are attracted to the idea of moving themselves up the food chain—profiting from owning some of the bits rather than from just delivering them. This is why cable companies, regional telephone companies, and consumer-electronics manufacturers are working with Hollywood studios and television and cable broadcasters. All kinds of alliances have formed.

There will also be a large number of mergers—some to create a bigger company within a business area, and others to combine companies in different businesses. Media companies are growing through mergers. Some phone companies are consolidating, moving into long-distance or local service, or buying cable companies. AT&T bought McCaw Cellular, a wireless communications company. Disney bought Capital Cities–ABC, the television network. Time Warner bought Turner Broadcasting. US West may have bought Continental Cablevision by the time you read this.

One of the lessons of the computer industry—as well as life—is that it's almost impossible to do everything well. IBM and DEC and other companies in the old computer industry tried offering everything— chips, software, systems of all sizes, and consulting. The diversification strategy made these companies vulnerable once the microprocessor and PC standards stepped up the pace of innovation. Over time their competitors who focused on specific areas did better. One competitor did great chips, another did great PC designs, yet another did great distribution and integration. Each successful new company picked a narrow slice of the pie and focused on it. Businesses that concentrate on a few core competencies will do the best in the coming era too.

Another lesson from the PC industry is that, when companies say they have an alliance, it can mean anything from a friendly pat on the back to joint efforts of real importance. Some of the warmest words are said between companies who are just patting each other on the back. Some of the significant alliances, such as the one between Intel and Microsoft, didn't receive much publicity when they were formed. In this world of complicated alliances, companies must be able to partner on some projects and compete vigorously on others. Few companies in the computer and communications industries are purely friends or purely foes.

Just because a company is big or has alliances is no assurance it will succeed. There's even more room for entrepreneurship than there has been in the past. Only a handful of the companies that made mainframe software managed the transition to personal computers. Most successes were startups run by people who were open to new possibilities. This will be true of interactive network successes too. For every large existing company that succeeds with a new application or service, ten startups will flourish and fifty more will flash into existence and momentary glory before slipping into obscurity.

This kind of activity is the hallmark of an evolving entrepreneurial market. Rapid innovation occurs on many fronts. Most of the innovation will be unsuccessful regardless of whether it's attempted by a large or a small company. Large companies tend to take fewer risks, but when they crash and burn, the combination of sheer ego and large-scale resources means that they dig a big crater. A startup usually fails without much notice. The good news is that people learn from both the successes and the failures, and the net result is rapid progress.

Nowhere is the benefit of a market-driven decision more apparent than in an unproven market. By letting the marketplace decide which companies and approaches win and lose, we can explore many paths simultaneously. When hundreds of companies risk different approaches to discover the kinds and levels of demand, society gets to the right solutions a lot faster than it would with any form of central planning. There is a lot of uncertainty in the communications revolution, but the marketplace will design a system that does what users want at low prices.

Governments can help ensure a strong competitive framework and should be willing although not overeager to intercede if the marketplace fails in some particular area. Governments can help determine the "rules of the road," the guidelines within which companies compete, but they shouldn't try to design or dictate the nature of the network because governments aren't very good at outguessing the competitive marketplace.

I hope other governments will follow the U.S. government's lead and deregulate their communications industries. The old policy in most countries, including the United States, was to create monopolies in many services that were perceived as essential. The assumption behind this approach was that a company wouldn't make the huge investments necessary to run telephone wires out to everyone, for example, unless it had the incentive of being the exclusive supplier. A set of rules drawn up by the government would bind the monopoly holder to act in the public interest with restricted but essentially guaranteed profit. The outcome in the U.S. was a very reliable telephone network with good service but limited innovation.

One area it's clear that government should stay out of is creating technology standards. Some people have suggested that governments set standards for networks to guarantee that they'll interoperate. In 1994 legislation was put before a subcommittee in the U.S. House of Representatives calling for all TV set-top boxes to be compatible. This sounded like a fine idea to the people who drafted the legislation. It would ensure that if someone invested in a set-top box, she could be confident it would work if she moved to another part of the country. For the sake of progress, it's fortunate that the legislation died.

Consumers, not regulators, should choose between compatibility and innovation. When the PC industry was new, many machines came and went. The Altair 8800 was superseded by the Apple I. Then came the Apple II, the original IBM PC, the Apple Macintosh, the IBM PC AT, the 386 and 486 PCs, Power Macintoshes, Pentium PCs, and Pentium Pros. Each of these machines was somewhat compatible with the others. All were able to share plain-text files, for instance. But there was also a lot of incompatibility because each successive computer generation provided fundamental breakthroughs the older systems didn't support.

Compatibility with earlier machines is a virtue in some cases. Both PC-compatibles and the Apple Macintosh provide some backwards compatibility with earlier models that used their machine architecture. However, the PC and the Mac are largely incompatible with each other. And at the time the PC was introduced, it wasn't compatible with IBM's earlier machines. Likewise, the Mac was incompatible with Apple's earlier machines.

The PC industry thrived precisely because technical standards were not regulated. If the U.S. Congress had decided in the early 1980s to "protect" consumers by requiring that all PCs adhere to the technical standards of mainframe computers, many innovations we take for granted today would never have come to pass. Mainframe manufacturers would have been the big winners, and consumers the big losers.

In the summer of 1996, the U.S. Federal Communications Commission (FCC) was considering a proposal backed by broadcasters and television manufacturers that would create a mandatory standard for broadcast television formats. The proposed standard would allow for higher-quality TV images while locking in compatibility with some of today's broadcast technology. The computer industry, which sees TVs and PCs coming together technologically, opposed the proposed standard because it would hamper innovation in the name of compatibility.

Computer technology is so dynamic that any company should be able to come out with whatever new product it wants and let the market decide if it has made the right set of trade-offs. As we saw in chapter 3, the marketplace will do that quite emphatically. Because a TV (or TV set-top box) connected to the Internet will be a computer in every sense, it stands to reason that its development will follow the same pattern of rapid innovation that has driven PC technology. It would be foolish to impose the constraint of government-dictated design on an unfinished invention.

I expect that standards-setting efforts will be made in many countries. It seems easy to legislate reasonable-sounding constraints, but if we don't watch out, those constraints will strangle innovation.

Creating a competitive communications market is complicated in countries where the regulated telecommunications monopolies are owned by the government. PTTs in those countries are investing in new network infrastructure, but when government organizations are involved, things often move slowly. The pace of investment and deregulation worldwide will increase in the next ten years because politicians are recognizing that this is critical for their countries' long-term international competitiveness. In many election campaigns, candidates' platforms will include an Internet plank advocating investment in PCs for schools and better Internet connections for the entire country. The politicization of Internet issues will make them more visible. The competition among nations trying either to take the lead in development or to make sure they don't fall behind is creating a positive dynamic.

Some national governments may fund pieces of their countries' information infrastructures. A government investment can help an information highway to be constructed sooner than it might be otherwise, but governments need to make sure they take advantage of technology already proven in other countries. Otherwise, a country might end up with an inferior project built by engineers who were out of touch with the rapid pace of technological development.

Something like this happened in Japan with the Hi-Vision high-definition television project. MITI, the powerful Ministry of International Trade and Industry, and NHK, the government-run TV broadcasting company, coordinated an effort among Japanese consumer electronics companies to build a new analog HDTV system. NHK committed to broadcasting shows a few hours a day in the new format. Unfortunately, the format was obsolete when it became clear that digital technology was superior. Japan will benefit from the Hi-Vision project's investment in developing high-definition cameras and displays, though.

Unless poor political decisions are made, I doubt that many of the developed countries will end up more than a year or two ahead of or behind the others in the development of an Internet-based communications infrastructure. But bad policy decisions will be made here and there. If we look back in ten years, we'll see a clear correlation between the amount of telecommunications reform in each country and the state of its information economy. There are enough countries involved in information technology that I'm sure the entire spectrum of different regulatory and deregulatory schemes will be tried.

We never got to see many trials of interactive television, but pioneering broadband networks such as the one planned for Singapore may help the entire industry understand what kinds of interactive broadband applications will have the most appeal. The potential downside in an effort such as Singapore's, of course, is that trailblazing, isolated systems run the risk of falling behind technologically because they miss out on the vital lessons of the global marketplace.

No matter how smart a single government or company is, in the long run it won't be able to match the innovation that is already resulting from feverish competition among the thousands of companies caught up in the new gold rush. The marketplace is the greatest decision-maker, smarter than any business or political leader, and it's hard at work guiding the development of the global communications standard—the Internet.

----------------------------->

12

CRITICAL ISSUES

T his is an exciting time in the Information Age. It is the very beginning. Almost everywhere I go, whether to speak to a group or have dinner with friends, questions come up about how information technology will change our lives. People want to understand how it will make the future different, whether it will make our lives better or worse.

It should be obvious by now that I'm an optimist about the impact of the new technology. It will enhance our leisure time and enrich our culture by expanding the distribution of information. It will help relieve pressures on urban areas by enabling people to work from home or remote-site offices. It will relieve pressures on natural resources because increasing numbers of products will take the form of bits rather than manufactured goods. It will give us more control over our lives, enabling us to tailor our experiences and the products we use to our interests. Citizens of the information society will enjoy new opportunities for productivity, learning, and entertainment. Countries that move boldly and in concert with each other will enjoy economic rewards. Whole new markets will emerge, and a myriad new opportunities for employment will be created.

For the past few hundred years every generation has found more efficient ways of getting work done, and the cumulative benefits have been enormous. The average person today enjoys a much better life than the nobility did a few centuries ago. It would be great to have a medieval king's land, but what about his lice? Medical advances alone have greatly increased life spans and improved standards of living.

In the first part of the twentieth century, Henry Ford *was* the automotive industry, but your car is superior to anything he ever drove. It's safer and more reliable, and it certainly has a better sound system. This pattern of improvement isn't going to change. Advancing productivity propels societies forward, and it's only a matter of time before the average person in a developed country will be "richer" in many ways than anyone is today.

Just because I'm optimistic doesn't mean that I don't have concerns about what's going to happen to all of us. Major changes always involve tradeoffs, and the benefits of the information society will carry costs. Societies are going to be asked to make hard choices about the universal availability of technology, investment in education, regulation, and the balance between individual privacy and community security. We'll confront tough new problems, only a few of which we can foresee. In some business sectors, dislocations will create a need for worker retraining. The availability of virtually free communications and computing will alter the relationships of nations and of socioeconomic groups within nations. The power and versatility of digital technology will raise new concerns about individual privacy, commercial confidentiality, and national security. There are equity issues that will have to be addressed because the information society should serve all of its citizens, not just the technically sophisticated and the economically privileged. I don't necessarily have the solutions to the many issues and problems we'll confront, but as I said at the beginning of this book, now is a good time for a broad debate to begin.

While it's important that we start thinking about the future, we should guard against the impulse to take hasty action. We can ask only the most general kinds of questions today, so it doesn't make sense to come up with detailed, specific recommendations and regulations. The pace of technological change is so fast that sometimes it seems as if the

----------------------------->

world will be completely different from one day to the next. It won't be, but we should prepare for change. We have a good number of years to observe the course of the coming revolution, and we should use that time to make intelligent rather than reflexive decisions.

Perhaps the most widespread and personal anxiety is, "How will I fit into the evolving economy?" People are worried that their jobs will become obsolete, that they won't be able to adapt to new ways of working, that their children will get into deadend industries that will close down, or that economic upheaval will create wholesale unemployment, especially among older workers. These are legitimate concerns. Entire professions and industries will fade. But new ones will flourish. These changes will be happening over the next two or three decades, which is fast by historical standards but probably no more disruptive than the pace at which the microprocessor revolution brought its changes to the workplace or deregulation and competition brought change to the airline, trucking, and banking industries.

Although the microprocessor and the personal computer it enabled have altered and even eliminated some jobs and companies, it's hard to find any large sector of the economy that has been negatively affected. Mainframe, minicomputer, and typewriter companies have downsized, but the computer industry as a whole has grown, with a substantial net increase in employment. As big computer companies such as IBM or DEC have laid people off, many of those workers have found employment elsewhere within the computer industry—usually at companies doing something related to PCs.

Outside the computer industry, it's also hard to find a whole business sector hurt by the PC. Some typesetters were displaced by the advent of desktop publishing applications, but many of them retrained. For every worker who lost a job there are several whose jobs desktop publishing created. All of the changes haven't always been good for all of the people, but as revolutions go, the one set in motion by the personal computer has been remarkably benign.

Some people worry that there are only a finite number of jobs in the world and that each time a job goes away someone is left stranded with no further purpose in life. Fortunately, this is not how the economy works. The economy is a vast interconnected system in which any resource that is freed up becomes available to another area of the economy that finds it valuable. Each time a job is made unnecessary, the person who was filling that job is freed to do something else. The net result is that more gets done, raising the overall standard of living in the long run. If there is a general downturn across the economy—a recession or a depression—there is a cyclical loss of jobs, but the shifts that have come about as a result of technology have tended, if anything, to create jobs.

In an evolving economy, job categories change constantly. At one time all telephone calls were made through an operator. When I was a child, long-distance calls from our house were made by dialing "0" and giving an operator the number, and when I was a teenager, many companies still employed in-house switchboard operators who routed calls by plugging cables into receptacles. Today there are comparatively few telephone operators even though the volume of calls is greater than ever. Automation has taken over.

Before the Industrial Revolution most people lived or worked on farms. Growing food was mankind's main preoccupation. If someone had predicted back then that within a couple of centuries only a tiny percentage of the population would be needed to produce food, all those farmers would have worried about what everyone would do for a living. The great majority of the 501 job categories recognized in 1990 by the U.S. Census Bureau didn't even exist fifty years earlier. Although we can't predict what the new job categories will be, I think that most of them will relate to unmet needs in education and social services and to leisure opportunities.

We do know that when the interactive network connects buyers and sellers directly, it will put pressure on people who are currently acting as middlemen. This is the same sort of pressure that mass merchants such as Wal-Mart, Price-Costco, and other companies with particularly efficient consumer-merchandising approaches have already put on more traditional stores. When Wal-Mart moves into a rural area, the merchants in the local towns feel the pinch. Some survive, some do not, but the net economic effect on the region is modest. We may regret the cultural ramifications, but warehouse stores and fast-food chains are thriving because consumers, who vote with their dollars, tend to support outlets that pass their productivity savings along in the form of lower prices.

Reducing the number of middlemen is another way of lowering costs. It will also cause economic shifts, but no faster than the rate at which shifts have been happening in retailing in the last decade. It will take many years for the Internet to be used so widely for shopping that there will be significantly fewer middlemen. There is plenty of time to prepare. The jobs those displaced middlemen change to might not even have been thought of yet. We'll have to wait and see what kinds of creative work the new economy devises. But as long as society needs help, there will definitely be plenty for everyone to do.

The broad benefits of advancing productivity are no solace for someone whose job is on the line. When a person has been trained for a job that's no longer needed, you can't just suggest he go out and learn something else. Adjustments aren't that simple or fast, but ultimately they are necessary. It isn't easy to prepare for the next century because it's almost impossible to guess the secondary effects of even the changes we can foresee, much less of those we can't. A hundred years ago people saw the automobile coming. It looked sure to make fortunes and just as sure to run over some jobs and industries. But specifics would have been hard to predict. You might have warned your friends at the Acme Buggy Whip Company to polish up their résumés and perhaps start learning a little something about engines, but would you have known to invest in real estate for strip malls?

More than ever, an education that emphasizes general problem-solving skills will be important. In a changing world, education is the best preparation for being able to adapt. As the economy shifts, people and societies who are appropriately educated will tend to do best. The premium that society pays for skills is going to climb, so my advice is to get a good formal education and then keep on learning. Acquire new interests and skills throughout your life.

A lot of people will be pushed out of their comfort zones, but that doesn't mean that what they already know won't still be valuable. It does mean that people and companies will have to be open to reinventing themselves—possibly more than once. Companies and governments can help train workers, but the individual must ultimately bear principal responsibility for his education.

A first step will be to come to terms with computers. Computers make almost everyone except children nervous before they understand them. First-time adult users worry that a single misstep will cause them to ruin the computer or lose everything stored in it. People do lose data, of course, but the damage is rarely irreversible. The industry has worked to make it harder to lose data and easier to recover from mistakes. Most programs have "Undo" commands that make it simple to try something and then quickly reverse it. Users become more confident as they see that making mistakes won't be catastrophic. And then they begin to experiment. PCs provide all kinds of opportunities for experimentation. The more experience people have with PCs, the better they understand what they can and can't do. Then PCs become tools instead of threats. Like a tractor or a sewing machine, a computer is a machine we can use to help us get certain tasks done more efficiently.

Another fear people express is that computers will be so "smart" they will take over and somehow outfox or do away with the need for human intelligence. Although I believe that eventually there will be programs that will re-create some elements of human intelligence, I don't think it's likely to happen in my lifetime. For decades computer scientists studying artificial intelligence have been trying to develop a computer with human understanding and common sense. Alan Turing suggested in 1950 what has come to be called the Turing Test: If you're able to carry on a conversation with a computer and another human, both hidden from your view, and you're uncertain about which is which, you have a truly intelligent machine.

So far every prediction about major advances in artificial intelligence has proved to be overly optimistic. Today even simple learning tasks are well beyond the world's most capable computer. When computers appear to be intelligent, it's because they've been specially programmed to handle some specific task in a straightforward way—like trying out billions of chess moves in order to play master-level chess. Computers such as Deep Blue, an IBM machine developed specifically to play chess, can sometimes even beat the world's best human players. But so what? A computer's chess prowess is not an approximation of human intelligence in any sense. It merely reflects the computer's ability to consider billions of different outcomes of a potential move.

-------------------------->

World chess champion Gary Kasparov beat Deep Blue in a 1996 match, but even Kasparov admitted he had a scare when the computer won the first game of the series. That was the only game the computer won, though. Kasparov came back with three wins and two draws to win the match. The contest was made out to be a "human versus machine" showdown, but in a real sense the contest was between a human— Kasparov—and an opposing committee of programmers and chess players who used a tool—Deep Blue—to carry out their carefully prede-termined strategies. Kasparov won that match, but he might not always win. When a machine finally prevails convincingly in chess, it shouldn't offend human dignity any more than the realization that a person with binoculars can see faraway objects better than a person without binocu-lars can. As long as computers are being programmed for specialized tasks, they aren't learning—and if they aren't learning, they pose no threat to humans.

I am troubled, however, by the possibility that in the very long run computers and software could achieve true intelligence. Predicting a timeframe is impossible because progress in artificial intelligence re-search is so incredibly slow. But once machines can really learn, they will be able to take over most things humans do today. This will raise issues of who is in control and what the whole purpose of our species is.

Of course, these worries aren't new. Mary Shelley's 1818 novel, *Frankenstein*, is perhaps the most famous story of an intelligent man-made creature that comes to control its creator. In 1942, when he was in his early twenties, Issac Asimov laid out his "Three Laws of Robotics," which he proposed as measures we enforce to ensure that intelligent machines won't exercise a dangerous level of independent judgment: "1. A robot may not injure a human being or, through inaction, allow a human being to come to harm. 2. A robot must obey orders given it by human beings except where such orders would conflict with the First Law. 3. A robot must protect its own existence as long as such protection does not conflict with the First or Second Law." More recently movies such as *Terminator* have aroused anxieties over the prospect of conflicts between people and intelligent machines.

In the nearer term I worry about the empowerment of terrorists. Technology is enabling small groups of people to do very destructive things. This is a frightening trend because defensive weaponry can't keep up with advances in offensive weaponry. So despite my general optimism, I'm not without worries about technology.

By and large, technology is a positive force that can help us solve even our most vexing potential problems. Two centuries ago the economist Thomas Malthus warned of the dangers of compounding population increases. Overpopulation remains a potentially ominous problem, given the effect of exponential growth, but recent trends are encouraging: Population growth slows as technology increases affluence and improves education. Environmental problems and resource shortages must be taken seriously too, but I think many doomsayers vastly underestimate the potential of technology to help us overcome these problems.

Frankly, I'm surprised by the pessimism many people feel about the future. The worldwide movement toward capitalism and democracy is encouraging, if uneven. Advances in medicine continue steadily, and biotechnology promises astounding breakthroughs that will greatly improve the human condition. So without in any way dismissing the seriousness of the world's problems or the plights of some societies, I have to say that pessimism about the future doesn't seem to be warranted.

In any event, the computer should be a source of optimism for the foreseeable future. The computer leverages human intelligence in a rather magical way. It would be wonderful if everybody—rich or poor, urban or rural, old or young—could have access to a personal computer, but they're still too expensive for most people. Before use of the global interactive network can become fully integrated into society, a computer hooked up to the Internet must be available to every citizen, not just the elite.

This doesn't mean that every citizen has to have an information appliance in his house. Once the majority of people have systems installed in their homes, those who don't can be accommodated with shared appliances at a library, a school, a post office, or a public kiosk. Keep in mind that the question of universal access arises only if the interactive network is immensely successful—more successful than many commentators expect it to be. Some of the same critics who complain that the network will be so popular it will cause all kinds of problems also complain dismissively that it won't be popular at all.

-------------------------->

I am convinced that we need to bring computers into our classrooms and libraries. When society decided a few generations ago that everyone should have access to books, it took governments, philanthropists, and companies all getting together to make it happen—and it took a long time. Societies need to enlist all the resources they can to make universal access happen all over again with computers. Microsoft is among many companies trying to play a role in getting computers into our schools and libraries. For example, we're working with a number of library systems in the U.S. and Canada to pilot the idea that anybody who walks in can sit down at a PC and use the Internet. There are some key questions that pilot studies will help answer. How much demand is there for library access to the Internet? Does free library access really get people onto computers and the net who wouldn't have used them otherwise? How much staff is needed to be sure the system is used but not misused? Since societies can't afford to give everybody a computer at home, providing them in schools and libraries is our best hope for broad access. It'll take a lot of companies and communities working together to make it happen.

The fully developed interactive network will be affordable—almost by definition. An expensive system that connected up a few big corporations and rich people simply wouldn't be the information highway. It would be the information private road. The Internet won't attract enough great content to thrive if only the most affluent 10 percent of society can get to it. There are fixed costs to authoring content, and a large audience is required to make it affordable. Advertising revenue won't support the Internet if a major part of the eligible market doesn't embrace the network. If too many people are priced out or if the network doesn't appeal to enough people, the price of service will be cut. In the long run the broadband information network is a mass phenomenon, or it is nothing.

A large part of the money you'll spend on network services will be money you already spend today for the same services in other forms. In the past you may have shifted money you spent from records to compact discs or from movie tickets to videotape rentals. When broadband interactive service eventually arrives in your neighborhood, your spending for videotape rentals will go to video-on-demand movies. And even when you have only a midband connection, you'll redirect part of what you spend now on print periodical subscriptions to interactive information services and communities. Most of the money that now goes to local telephone service, long-distance service, and cable television will be available to spend on the network.

Access to government information, medical advice, bulletin boards, and some educational material will be free. Once people are on the network, they'll enjoy full egalitarian access to vital on-line resources. Within twenty years, as commerce, education, and broad-scale communications services move onto the network, an individual's ability to be part of mainstream society will depend, at least in part, on his or her using the network. Society will then have to decide how to subsidize broad access so that all users will be equal, both geographically and socioeconomically.

Education is not the entire answer to the challenges presented by the Information Age, but it is part of the answer, just as education is part of the answer to a range of society's problems. H.G. Wells, who was as imaginative and forward-looking as any futurist, summed it up back in 1920. "Human history," Wells said, "becomes more and more a race between education and catastrophe." Education is society's great leveler, and any improvement in education goes a long way toward equalizing opportunity. Part of the beauty of the electronic world is that the extra cost of letting additional people use educational material is basically zero.

Your education in personal computers can be informal. My fascination began with game playing, as years later Warren Buffett's did. My dad got hooked when he used a computer to help him prepare his taxes. If computers seem intimidating to you, why not try a similar approach? Find something a personal computer does that will make your life easier or more fun, and latch on to that as a way of getting acquainted with the computer. Write a screenplay. Join a discussion group. Do your banking and bill paying. Help your child with her homework. It's worth making the effort to establish a level of comfort with computers. If you give them a chance, you'll most likely be won over. If personal computing still seems too hard or confusing, it doesn't mean that you aren't smart enough. It means that the information industry still has to make computers easier to use.

----------------------------➤

The younger you are, the more important it is that you adapt. If you're fifty or older today, you may be out of the workforce before you'll need to learn to use a computer—although I think that if you don't learn you'll be missing out on the chance for some meaningful experiences. But if you're twenty-five today and not comfortable with computers, you risk being ineffective in almost any kind of work you pursue.

Ultimately the interactive network isn't for my generation or the one before me. It's for future generations. The kids who have grown up with PCs in the last decade, and the kids who will grow up with the network in the next, will push the technology to its limits.

We have to pay particular attention to correcting the gender imbalance. When I was young, it seemed that only boys were encouraged to mess around with computers. Girls are far more active with computers today than they were two decades ago, but there are still many fewer women than men in technical careers. By making sure that girls as well as boys get comfortable with computers at an early age, we can ensure that both sexes enjoy their rightful share of the good jobs that are available to people with computer expertise.

My own experience as a child and the experience of my friends raising children today is that, once a child is exposed to computing, he or she is hooked. But we have to create the opportunity for that exposure. Schools should have low-cost access to computers connected to the interactive network, and teachers need to become comfortable with the new tools.

One of the wonderful things about the interactive network is that virtual equity can be achieved much more easily than real-world equity. It would take a massive amount of money to give every grammar school in every poor area the same library resources as the schools in Beverly Hills. But when you put schools on-line, they all get the same access to information, wherever it might be stored. We are all created equal in the virtual world, and we can use this equality to help address some of the sociological problems that society has yet to solve in the physical world. The network won't eliminate barriers of inequality and prejudice, but it will be a powerful force in that direction.

The question of how to price intellectual property such as entertainment and educational materials is fascinating. Economists understand a lot about how the pricing of classical manufactured goods works. They can show how rational pricing should reflect cost structure in a very direct way. In a market with multiple competing qualified manufacturers, prices tend to drop to the marginal cost of making one more of whatever the manufacturers are selling. But this model doesn't work when it's applied to intellectual property.

Every basic economics course focuses on the curves of supply and demand, which intersect at the price appropriate for a product. But supply and demand economics gets into trouble when it comes to intellectual property because the ordinary rules that have to do with manufacturing costs and pricing don't apply. Typically there are huge up-front development costs for intellectual property. These fixed costs are the same regardless of whether one copy or a million copies of the work are sold. George Lucas's next movie in the *Star Wars* series will cost millions to make regardless of how many people end up paying to see it in theaters. Once the up-front costs of creating intellectual property are met, it's usually relatively inexpensive to manufacture additional copies.

Pharmaceuticals are in many respects a form of intellectual property. When you buy a new medicine, you're paying mostly for what the drug company spent for research, development, and testing. Even if the marginal cost of making each pill is minimal, the pharmaceutical company may still have to charge quite a bit for each one, especially if the market isn't that big. The revenue from the average patient has to cover a sufficient share of the development expenses and generate enough profit to make investors glad they took the substantial financial risks involved in developing the new drug. When a poor country wants the medicine, the manufacturer faces a moral dilemma. If the pharmaceutical company doesn't waive or drastically reduce its patent-licensing fees, the medicine simply won't be available to poor countries. As a result, prices for drugs vary greatly from country to country and tend to discriminate against poor people in rich countries except where governments cover medical costs.

One possible solution, a scheme whereby a rich person pays more to buy a new medicine—or to see a movie or read a book—may seem inequitable; however, this solution is essentially equivalent to a system already in place today: progressive taxation. Through the income tax and other taxes, people with high incomes pay more for roads, schools, the army, and every other government facility than the average person does. It cost me more than $100 million one recent year to get those services because I paid a significant capital gains tax after selling some Microsoft shares. I have no complaint because I recognize that sometimes it's fair to have the same services provided to different people at vastly different prices.

Eventually the pricing of network access may be set politically rather than in the marketplace. It's going to be expensive to enfranchise people in remote locations because the cost of bringing wiring to far-flung homes and even small communities is very high. Companies may not be eager to make the necessary investment, and the geographically disenfranchised may not be in a position to make the investment on their own behalf. Heated debate is under way about whether the government should subsidize connections to rural areas or impose regulations that cause urban users to subsidize rural ones. The "universal service" doctrine was formulated to provide a rationale for subsidizing rural mail, phone, and electrical services in the United States. Universal service says that there should be a single price for the delivery of a letter, for a phone call, or for electrical power regardless of where you live. The single price applies even though it's more expensive to deliver services in rural areas, where homes and businesses are farther apart than in areas of concentrated population.

There has been no equivalent policy for the delivery of newspapers or radio or television reception. Nevertheless, these services are widely available, so clearly under some circumstances government intervention isn't necessary to ensure high availability. The U.S. Postal Service was founded as part of the government on the assumption that government control was the only way to provide truly universal mail service. UPS and Federal Express might disagree on this point, however, because they've managed to provide broad coverage and make money. The debate over whether, or to what degree, government needs to be involved to guarantee broad access to the network is certain to rage for many years.

The network will let people who live in remote places consult, collaborate, and be involved with people in the rest of the world. Because many people will find the combination of rural lifestyle and urban information attractive, network companies will have an incentive to run fiber-optic lines to high-income remote areas. It's likely that some states or communities or even private real estate developers will promote their areas by providing great connectivity. Interesting rural communities with high marks for quality of life will deliberately set out to attract a new class of sophisticated urban citizen. On balance, though, urban areas will tend to get their connections before rural areas.

The network will spread information and opportunity across borders to developing nations too. Cheap global communications can bring people anywhere into the mainstream of the world economy. Knowledge workers in industrialized countries will, in a sense, face new competition—just as some manufacturing workers in industrialized countries have experienced competition from developing nations over the past decade. An English-speaking Ph.D. in China will be able to bid against colleagues in London for consulting work. The network will be a powerful force for international trade in intellectual goods and services, just as the availability of relatively inexpensive air cargo and containerized shipping helped propel international trade in physical goods.

The net effect will be a wealthier world, which should be stabilizing. Developed nations and workers in those nations are likely to maintain a sizable economic lead, but the gap between the have and the have-not nations will diminish—great news for the countries that are behind economically today. Starting out behind is sometimes an advantage. Some developing countries will never pass through the "industrialization" stage with its attendant problems. They'll move directly into the Information Age. Late adopters can skip steps and avoid the mistakes of the trailblazers. Europe didn't adopt television until several years after the United States. The result was higher picture quality for Europeans because by the time Europe set its standard a better choice was available. Europeans have enjoyed better-looking television pictures for decades.

Telephone systems are another example of how starting late can be an advantage. In Africa, Asia, Latin America, and other developing regions, cellular telephone service is spreading rapidly because it doesn't require copper wire infrastructure. Many people in the cellular industry predict that improvements in technology will mean that these areas may never get a conventional copper wire–based telephone system. These

- >

countries will never have to cut down a million trees for telephone poles or string a hundred thousand miles of telephone lines only to rip them all down and bury the entire network a few decades later. The wireless telephone system will be their first telephone system, and they'll get increasingly better cellular systems wherever they can't afford broadband connections.

The presence of advanced communications systems promises to make countries more alike and reduce the importance of national boundaries. The fax machine, the portable videocamera, and Cable News Network are among the forces that brought about the end of communist regimes and the Cold War because they enabled news to pass both ways through the Iron Curtain. Most sites on the World Wide Web are in English so far, which confers economic and entertainment benefits on people around the world who speak English. English-speaking people will enjoy this advantage until a great deal more content is posted in a variety of languages—or until software does a first-rate job of translating text on the fly.

The new access to information can draw people together by increasing their understanding of other cultures. But commercial satellite broadcasts to countries such as China and Iran offer citizens glimpses of the outside world that are not necessarily sanctioned by their governments. Some governments are afraid that such exposure will cause discontent and worse, a "revolution of expectations" when disenfranchised people get enough information about another lifestyle to contrast it with their own. Within individual societies, the balance between traditional and modern experiences is bound to shift as people use the network to expose themselves to a greater range of possibilities. Some cultures may feel under assault as people pay greater attention to global issues and cultures and less to their traditional local ones.

"The fact that the same ad can appeal to someone in a New York apartment and on an Iowa farm and in an African village does not prove these situations are alike," commented Bill McKibben, a critic of what he sees as television's tendency to override local diversity with homogenized common experiences. "It is merely evidence that the people living in them have a few feelings in common, and it is these barest, most minimal commonalties that are the content of the global village."

Yet if people want to watch the ad, or the program the ad supports, should they be denied that privilege? This is a political question for every country to answer individually.

American popular culture is so potent that outside the United States some countries now try to ration exposure to it. They hope to guarantee the viability of domestic content producers by permitting only a certain number of hours of foreign television to be aired each week, for instance. In Europe, the availability of satellite and cable-delivered programming has made it harder for governments to control what people watch. The Internet is going to break down boundaries and may promote a world culture, or at least a greater sharing of cultural activities and values. But the network will also make it easy for people who are deeply involved in their own ethnic communities at home or abroad to reach out to other people who share their preoccupations no matter where they are. This may strengthen cultural diversity and counter the tendency toward a single, homogenized world culture. It's hard to predict what the net effect will be—a strengthening or a weakening of local cultural values.

If people do gravitate toward their own parochial interests and withdraw from the broader world—if weight lifters communicate only with other weight lifters, and Latvians choose to read only Latvian newspapers—there is a risk that common experiences and values will fall away. Such xenophobia would have the effect of fragmenting societies. I doubt that this will happen because I think people want a sense of belonging to multiple communities, including a world community. When Americans share national experiences, it's usually because we're witnessing events all at the same time on television—whether it's the tragedy of the *Challenger* blowing up after liftoff, the Super Bowl, an inauguration, coverage of the Gulf War, or the O. J. Simpson car chase. We're "together" at those moments. The whole world was watching when the Berlin Wall came down, and the whole world will increasingly have global experiences together. Unfortunately, with a few exceptions such as the Olympics and man walking on the moon for the first time, most global experiences have tended to be experiences of negative phenomena: natural disasters, war, and other kinds of strife.

Another concern people raise is that multimedia entertainment will be so easy to get and so compelling that some of us will use the system too much for our own good. This could become a problem when virtual reality technology gets really good and becomes commonplace.

One day a virtual reality game will let you enter a virtual bar and make eye contact with "someone special," who will note your interest and come over to engage you in conversation. You'll talk, impressing this new friend with your charm and wit. Perhaps the two of you will decide, then and there, to go to Paris. Whoosh! You'll be in Paris, gazing together at the stained glass windows of Notre Dame. "Have you ever ridden the Star Ferry in Hong Kong?" you might ask your friend. Whoosh! VR will certainly be more engrossing than video games have ever been, and more addictive.

If you found yourself escaping into such an attractive world too often, or for too long, and you started to worry about it, you could impose some discipline on yourself by telling the system, "No matter what password I try, don't let me play any more than half an hour of games a day." This would be a little speed bump, a warning to slow your involvement with something you've found a little too appealing. It would serve the same purpose as a photo of some very overweight people you might post on your refrigerator to keep you from snacking.

Speed bumps help a lot with behavior that tends to generate day-after regrets. If somebody wants to spend his or her free time window gazing in a simulation of Notre Dame, or visiting with a synthetic friend in a make-believe bar, that person is exercising his or her freedom. Today a lot of people spend several hours a day with a television on. To the extent that we can replace some of that passive entertainment with interactive entertainment, viewers may be better off. Frankly, I'm not too concerned about the world's whiling away its hours on the Internet. At worst, I expect that it will be like playing too many video games or overeating.

A more serious concern than individual overindulgence is the vulnerability that could result from society's growing, heavy reliance on the network.

The global interactive network and the computer-based machines connected to it will constitute society's new playground, new workplace, and new classroom. It will replace physical tender. It will subsume most existing forms of communication. It will be our photo album, our diary, our boom box. This versatility will be the strength of the network, but it will also mean we'll become extraordinarily reliant on it.

That much dependence can be dangerous. During the New York City blackouts in 1965 and 1977, millions of people were in trouble—at least for a few hours—because of their dependence on electricity. They counted on electric power for light, heat, transport, and security. When electricity failed, people were trapped in elevators, traffic lights stopped working, and electric water pumps quit.

A complete failure of the network is worth worrying about. Because the system will be thoroughly decentralized, any single outage is unlikely to have a widespread effect. If an individual server fails, it will be replaced and its data restored. But the system could be susceptible to assault. As the network becomes more essential to us, we'll have to design in more redundancy.

One area of vulnerability is the system's reliance on cryptography—the mathematical locks that keep information safe. None of the protection systems that exist today, whether steering-wheel locks or steel vaults, are completely fail-safe. The best we can do is make it as difficult as possible for somebody to break a security device or get inside. Despite popular impressions to the contrary, computer security has a very good record. Computers are capable of protecting information in such a way that even the smartest hackers can't get at it readily unless somebody who is entrusted with the information makes a mistake or is corrupt. Sloppiness is the main cause of computer security breaches. On the network there will be mistakes, and sometimes too much information will get passed along. Somebody will issue digital concert tickets that prove to be forgeable, and too many people will show up at the concert. Whenever this sort of thing happens, the system will have to be reworked and laws may have to be revised.

The calamitous potential for forgery of digital money was demonstrated in 1996 when a Japanese company partly owned by Mitsubishi reported a loss of 55 billion yen ($588 million) because of counterfeit magnetic pachinko cards. Pachinko is a passion in Japan, where more than 18,000 smoky parlors are devoted to the game, which is a cousin of pinball. According to Japanese news reports, organized crime syndicates are behind the counterfeiting of the prepaid cards, which players use to pay for pachinko games—and which can be redeemed indirectly for cash.

Because both the Internet's privacy and the security of digital money depend on encryption, a breakthrough in mathematics or computer science that defeats the cryptographic system could be a disaster. It was sobering to me when certain kinds of public key encryption were revealed to be vulnerable to a "timing attack," a development I discussed in chapter 4. The obvious mathematical breakthrough that would defeat our public key encryption would be the development of an easy way to factor large numbers. Any person or organization who had this capability could counterfeit money, penetrate any personal, corporate, or government file, and possibly even undermine the security of nations, which is why we have to be so careful in designing the network's encryption system. We have to ensure that if any particular encryption technique proves fallible, there is a way to make an immediate transition to an alternative technique. There's a little bit of inventing still to be done before we have that perfected. The implications of Moore's Law make it particularly hard to guarantee security for information you want to be kept confidential for a decade or more. More powerful computers can lead to the breakthroughs that can violate our security.

Loss of privacy is another major worry where the network is concerned. A great deal of information is already being gathered about each of us, by private companies as well as by government agencies, and we often have no idea how it's used or whether it's accurate. Census Bureau statistics contain significant amounts of detail about each of us. Medical records, driving records, library records, school records, court records, credit histories, tax records, financial records, employment reviews, and charge-card bills all profile us. The fact that you call a lot of motorcycle shops and the surmise that you might therefore be susceptible to motorcycle advertising is commercial information a telephone company theoretically could sell. Information about us is routinely compiled into direct marketing mailing lists and credit reports. Errors and abuses in credit reports have already stimulated legislation regulating the use of databases that contain information about us. In the United States, you're entitled to see certain kinds of information that's stored about you, and you may have the right to be notified when someone looks at it. The scattered nature of information protects your privacy in an informal way, but when the repositories of information are all connected together on the network, it will be possible to use computers to correlate the scattered information. Credit data could be linked with employment records and sales transaction records to construct an intrusively accurate picture of your personal activities.

As more business is transacted by means of the network and the amount of information stored there accumulates, governments will set policies regarding privacy and access to information. Software will administer the policies, ensuring that a doctor doesn't get access to a patient's tax records, that an IRS auditor can't look at a taxpayer's scholastic record, that a teacher isn't permitted to browse a student's medical record. The potential problem is not the mere existence of information. It's the abuse that makes me worry.

We now allow a life insurance company to examine our medical records before it determines whether it wants to insure our mortality. An insurance company may also want to know whether we indulge in any dangerous pastimes, such as hang gliding, smoking, or stock car racing. Should an insurer's computer be allowed to cruise the network for records of our purchases to see whether there are any that might indicate risky behavior on our part? Should a prospective employer's computer be allowed to examine our communications or entertainment records to develop a psychological profile? How much information should a federal, state, or city agency be allowed to see? What should a potential landlord be able to learn about you? What information should a potential spouse have access to? We'll need to define both the legal and the practical limits of privacy.

These privacy issues revolve around the possibility that somebody else is keeping track of information about you. But the network will also make it possible for you to keep track of your own whereabouts—to lead what we might call a "documented life."

Your wallet PC will be able to keep audio, time, location, and eventually even video records of everything that happens to you. It will be able to record every word you say and every word said to you, as well as your body temperature, your blood pressure, the barometric pressure, and a variety of other data about you and your surroundings. It will be able to

-----------------------------➤

track your interactions with the network—all of the commands you issue, the messages you send, and the people you call or who call you. The resulting record will be the ultimate diary and autobiography, if you want one. If nothing else, you'd know exactly when and where you took a photograph when you organized your family's digital photo album.

The technology to do this is not just a remote possibility. It should soon be possible to compress the human voice down to a few thousand bits of digital information per second, which means that an hour of conversation will be converted into about 1 megabyte of digital data. Small tapes used for backing up computer hard disks already store 10 gigabytes or more of data, enough to record about 10,000 hours of compressed audio—more than a year's worth. Tapes for new generations of digital VCRs will hold more than 100 gigabytes, which means that a single tape costing a few dollars could hold recordings of all the conversations an individual has over the course of a decade or possibly even a lifetime—depending on how talkative he is. These numbers are based on today's capacities; in the future, storage will be much more capacious and cheaper. Audio recording will be easy, and within a few years a video recording will be possible too.

I find the prospect of documented lives chilling, but some people will like the idea. For one thing, a documented life can be a good defense. We can think of the wallet PC as an alibi machine because encrypted digital signatures will guarantee an unforgeable alibi against false accusations. If someone ever accused you of something, you could retort: "Hey, buddy, I have a documented life. Those bits are stored away. I can play back anything I've ever said. So don't play games with me." Of course, if you were guilty of something, there would be a record of it. There would also be a record of any tampering you tried. Richard Nixon's taping of conversations in the White House—and then the suspicion that he tried to alter those tapes—contributed to his undoing. He chose to have a documented political life and came to regret it.

The Rodney King case showed both the evidentiary power of videotape and its limits. Before long every police car or individual policeman may be equipped with a digital video camera, with unforgeable time and location stamps. The public may insist that the police record themselves in the course of their work. And the police could be all for it, to guard against claims of brutality or abuse on the one hand and as an aid in gathering better evidence on the other. Some police forces are already video-recording all arrests. This sort of record won't affect just police work. Medical malpractice insurance might be cheaper or available only for doctors who record surgical procedures or even office visits. Bus, taxi, and trucking companies have an obvious interest in the performance of their drivers. Some transportation companies have already installed equipment to record mileage and average speed. I can imagine proposals that every automobile, including yours and mine, be outfitted not only with a recorder but also with a transmitter that identifies the car and its location—a future digital license plate. After all, airplanes have "black box" recorders today, and once the cost drops, there's no reason they shouldn't also be in our cars. If a car was reported stolen, its location would be known immediately. After a hit-and-run accident or a drive-by shooting, a judge could authorize a query: "What vehicles were in the following two-block area during this thirty-minute period?" The black box could record your speed and location, which would allow for the perfect enforcement of speeding laws. I would vote against that.

In a world that's increasingly instrumented, we could reach the point where cameras recorded most of what goes on in public. Video cameras in public places are already relatively commonplace. They perch, often concealed, around banks, airports, automatic-teller machines, hospitals, freeways, stores, and hotel and office-building lobbies and elevators.

The prospect of so many cameras, always watching, might have distressed us fifty years ago, as it did George Orwell. But today it seems relatively unremarkable. There are neighborhoods in the United States and Europe where citizens are welcoming these cameras above streets and parking lots. In Monaco, street crime has been virtually eliminated because hundreds of video cameras have been placed around the tiny principality. (Monaco is small enough, 370 acres—150 hectares— that a few hundred cameras can pretty much cover it all.) Many parents would welcome cameras around schoolyards to discourage or help apprehend drug dealers, child molesters, and even playground bullies. Every city streetlight represents a substantial investment by a community in public safety. In a few years it will require only a relatively modest additional sum to add and operate cameras with connections to the information

----------------------------➤

network. Within a decade, computers will be able to scan video records very inexpensively looking for a particular person or activity. I can easily imagine proposals that virtually every pole supporting a streetlight should also have one or more cameras. The images from these cameras might be accessed only in the event of a crime, and even then possibly only under court order. Some people might argue that every image from every camera should be available for viewing by everyone at any time. This raises serious privacy questions in my mind, but advocates might argue that it's appropriate if the cameras are only in public places.

Almost everyone is willing to accept some restrictions in exchange for a sense of security. It's a question of balance. From a historical perspective, people living in the Western democracies already enjoy a degree of privacy and personal freedom unprecedented in human history. If ubiquitous cameras tied into the network should prove to reduce serious crime dramatically in test communities, a real debate would begin over whether people fear surveillance more or less than they fear crime. It's difficult to imagine a government-sanctioned experiment along these lines in the United States because of the privacy issues it raises and the likelihood of constitutional challenges. However, opinion can change. It might take only a few more incidents within the borders of the United States like the bombings in Oklahoma City and Atlanta for attitudes toward strong privacy protection to shift. What today seems like digital Big Brother might one day become the norm if the alternative is being left at the mercy of terrorists and criminals. I'm not advocating a position, just pointing out that it's a political decision—and one over which there's likely to be a great deal of debate in the years ahead.

At the same time that technology is making it easier to create video records, it's also making it possible to keep all of your personal documents and messages totally private. Encryption-technology software, which anyone can download from the Internet, can transform a PC into a virtually unbreakable code machine. As the network is deployed, security services will be applied to all forms of digital information—phone calls, files, databases, you name it. As long as you protect the password, the information stored on your computer will be held under the strongest lock and key that has ever existed. This security allows for the greatest degree of information privacy any individual has ever had.

Some people in some United States government agencies are opposed to this encryption capability because it reduces their ability to gather information. Unfortunately for them, the technology can't be stopped. The National Security Agency is a part of the U.S. government defense and intelligence community that protects secret communications and decrypts foreign communications to gather intelligence data. The NSA wants to keep it illegal to send software containing advanced encryption capabilities outside the United States. The FBI wants to keep such exports illegal too, but for a different reason. The FBI believes that strong encryption systems won't be developed domestically, even if they're legal, if they can't also be sold in foreign markets. But good encryption software is already available throughout the world, and any PC can run it. No policy decision will be able to restore the decrypting capabilities governments have had in the past.

Today's laws that prevent the export of software with good encryption capability could harm U.S. software and hardware companies. The restrictions give foreign companies an advantage over U.S. competitors because they're allowed to build stronger security into their products. American hardware and software companies almost unanimously agree that the current encryption export restrictions don't work.

Each media advance has had a substantial effect on how people and governments interact. The printing press and later mass-circulation newspapers changed the nature of political debate. Radio and then television enabled government leaders to talk directly and intimately with the populace. Similarly, the interactive network will have its own influence on politics. For the first time politicians will be able to see immediate representative surveys of public opinion. Constituents will be able to tap out a quick letter to their congressional representative and find out what her positions are and how she's voting on this or that issue. Voters will be able to cast their ballots from home or their wallet PCs with less risk of miscounts or fraud.

Even if the model of political decision-making doesn't change explicitly, the network will bestow power on groups of citizens who want to organize to promote causes or candidates. This could lead to an increased number of special-interest groups and even political parties. Today organizing a political movement around a particular issue requires an immense amount of coordination. How do you find the people who share your view? How do you communicate with them and motivate them? Telephones and fax machines are great for connecting people one-on-one—but only if you know whom to call. Television lets one person reach millions, but it's expensive and wasteful if most viewers aren't interested in the issue.

Political organizations require thousands of hours of volunteer time. Envelopes have to be stuffed for direct mail appeals, and volunteers have to go out and contact people by whatever means possible. Only a few urgent issues, the environment being one, are potent and universal enough to overcome the difficulties involved in recruiting enough volunteers to operate an effective political organization.

The interactive network makes all communication easier. Bulletin boards and other on-line forums allow people to be in touch one-to-one, or one-to-many, or many-to-many, in strikingly efficient ways. People with similar interests can meet on-line and organize without any physical overhead. It will become so easy to organize a political movement that no cause will be too small or its constituents too scattered. Web sites became a significant focus for candidates and political action groups in the 1996 U.S. national elections, and eventually the network will become a primary conduit for political discourse.

Referendum voting is already used in the United States for specific issues at the state level. For logistical reasons these ballot propositions can go before the people only when a major election is already taking place. The Internet, however, could allow such votes to be scheduled far more frequently because they would cost very little to conduct.

Someone will doubtless propose total "direct democracy," having all issues put to a vote, instead of our current representative democracy. Personally, I don't think direct voting would be a good way to run a government. There is a place in governance for representatives—middlemen—to add value. They are the ones whose job it is to take the time to understand all the nuances of complicated issues. And politics involves compromise, which is nearly impossible without a relatively small number of representatives making decisions on behalf of the people who elected them. The art of management—whether of a society or a company—revolves around making informed choices about the allocation of resources. It's the job of full-time policymakers to develop expertise. This enables the best of them to devise and embrace nonobvious solutions that voters in a direct democracy might not even entertain because they might not understand the trade-offs necessary for long-term success.

Like all middlemen in the new electronic world, political representatives will have to justify their roles. The network will put the spotlight on them as never before. Instead of being given just photos and sound bites, voters will be able to get a much more direct sense of what their representatives are doing and how they're voting. It won't be long before a senator receives a million pieces of e-mail on a topic or has his beeper announce the results of a real-time opinion poll of his constituents.

As the Internet has grown in popularity around the world, the freedom it confers on anyone to distribute any kind of information almost anywhere has alarmed some people and governments. Libel, pornography, and copyright violations are widespread. Information about dangerous technologies is there for the taking. Political views that some people don't think anyone should have the right to express are voiced.

There are those who believe that the Internet is getting out of hand and needs to be reined in. This is a dangerous overreaction. The Internet is precious. It's the first medium that allows worldwide publication of information at essentially no marginal cost. If governments squeeze it too hard, it will suffocate.

At the other extreme there are people who insist that interactive networks should be wide open—that free cyberspace should be a self-governed world apart, in which copyright, libel, pornography, and confidentiality laws do not apply. This is naïve. The Web is mainstream now, and its days as a lawless backwater are over.

What's needed is balance. We must let the Internet be open even while we shield our children—and people who want to be shielded—from abuse. Material that is legal in libraries and bookstores must be legal on-line too, just as material that is libelous in print form should have the same status in cyberspace. We don't want to cripple the Internet, nor do we want children to have access to pornography or terrorism handbooks.

The U.S. Congress failed to strike a reasonable balance in 1996 when it passed the Communications Decency Act, which was so restrictive that it could have made it a felony to use the Internet to communicate detailed information about birth control, AIDS prevention, or how to get an abortion. A three-judge panel for the U.S. District Court in Philadelphia ruled the act an unconstitutional abridgement of free speech. "The Internet may be fairly regarded as a never-ending world-wide conversation," U.S. District Judge Stewart Dalzell wrote. "The government may not . . . interrupt that conversation. As the most participatory form of mass speech yet developed, the Internet deserves the highest protection from governmental intrusion." The government was appealing Judge Dalzell's ruling to the Supreme Court when this book went to press.

The U.S. Congress certainly isn't alone in trying to impose restrictions. The Chinese government requires users of the Internet and electronic mail to register and tries to restrict political expression broadly in the name of security and social stability. I don't approve of China's extreme position, but the truth is that most countries are sensitive to the distribution of one kind of information or another. The United Kingdom is keen on regulating the Internet with regard to state secrets and personal attacks. France has a proud heritage of press freedom, but a court banned a book on the health history of former French president Francois Mitterrand. Interestingly, it was legal for the book to be republished electronically on the World Wide Web, which it quickly was. Had it been illegal to publish it on the Web in France, the book's content could have been published on a Web server outside France and beyond the reach of French law—but not beyond the reach of French citizens browsing the Web. This is precisely the quandary Germany found itself in. Germany has strict laws against neo-Nazi propaganda, but German citizens browsing the Web had access to neo-Nazi viewpoints on a Web server in Canada, where the propaganda was posted legally. Keeping information out of a country is getting harder and harder.

As I've already noted, in our wired age it isn't easy to keep information inside national borders either. It used to be a lot easier, and governments have long had an interest in it for economic and military as well as political reasons. In the seventeenth century, it was a capital offense to disclose or try to discover the details of a process that used water power to twist raw silk into thread in a mill in Piedmont, Italy. In the days of Ivan the Terrible, smuggling a live sable out of Russia was punishable by beheading because Russia wanted to maintain its world monopoly on sable pelts. The information about how to "make" a sable pelt was (and is) locked into the sable's genetic code. The only way to export this information was to export live sables. Taking this live genetic information out of Russia is still a crime, by the way, although you wouldn't be beheaded for it now.

When it comes to the problem of keeping objectionable material from showing up on home PCs, part of the problem is that cyberspace doesn't have the gatekeepers we've come to rely on in the physical world. Society asks store clerks not to sell pornography, cigarettes, or alcohol to minors. Society asks pharmacists not to sell potent drugs to people without prescriptions. The clerk and the pharmacist serve as gatekeepers, although not always perfect ones. The World Wide Web has had no gatekeepers. It has been more like an unsupervised vending machine that offers almost anything to anyone—and usually for free. In what passes for gatekeeping on the Web, a site featuring adult material may display a statement that says something to the effect of "Warning! You must be twenty-one years of age to view the material on this site. To continue, click the 'I Am an Adult' button."

Some critics have suggested that communications companies be made gatekeepers, charged with filtering the content of what they carry. This idea would put companies in the business of censoring all communication. It's entirely unworkable, for one thing because the volume of communicated information is way too large. This idea is no more feasible or desirable than asking a telephone company to monitor and accept legal responsibility for everything that's spoken or transmitted on its telephone wires.

Politicians are wrestling with the question of when an on-line service should be treated as a common carrier and when it should be treated as a publisher. Telephone companies are legally considered common carriers. They transport messages without assuming any responsibility for them. If an obscene caller bothers you, the telephone company will cooperate with the police but nobody thinks it's the phone company's fault. Magazines and newspapers, on the other hand, are publishers.

They are legally responsible for the content they publish, and they can be sued for libel. They also have a strong interest in maintaining their reputation for editorial integrity because it's an important part of their business. Any responsible newspaper checks very carefully before making a previously unpublished allegation about someone—in part because it doesn't want a libel suit but also because publishing an inaccuracy would hurt its reputation for probity.

On-line services tend to function simultaneously as common carriers and publishers, which is where the problem lies. When a service acts as a publisher, offering content it has acquired, authored, or edited, it makes sense that it take responsibility for what it publishes. The rules of libel and a concern for editorial reputation should apply. But we also expect a service to deliver our e-mail like a common carrier without examining or taking responsibility for its contents. Likewise, the service's chat lines, bulletin boards, and forums that encourage users to interact without editorial supervision are a new means of communication that shouldn't be treated the same way material published by the service is. The stakes are high. If network providers are to be treated entirely as publishers, they'll have to monitor and preapprove the content of all the information they transmit. This could create an unwelcome atmosphere of censorship and curtail the spontaneous exchange so important in the electronic world.

Ideally, the industry will develop some standards on its own so that, when you visit a Web site or newsgroup, you'll get an indication of whether the "publisher" has looked it over, edited the content, and stands behind the content—or not. The question, of course, will be what standards? And who will oversee them? A bulletin board for lesbians shouldn't be forced to accept antilesbian comments, nor should a bulletin board about some product be overwhelmed by messages from a competitor. It would be a shame to have to keep children away from all Web sites or newsgroups, but it would also be unrealistic, and possibly an abridgment of free expression rights, to force all such sites to undergo review by someone willing to accept liability for everything they contain.

Perhaps the most workable approach is to require that Web pages and other on-line content be rated systematically. Web browsers and other applications are being given the ability to refuse connections to sites that don't have appropriate ratings from an authorized ratings organization. The Internet user—or the user's parent—can determine which rating service should be used and which of its standards should be applied. Parents and other users will be able to choose from a variety of rating services and approaches. One of these approaches is an initiative called PICS, for Platform for Internet Content Selection, that enables groups and companies worldwide to set their own standards for acceptable content. PICs is not a rating service, but rather a framework that makes it easy for organizations of all kinds to create their own ratings systems.

This won't make for perfect gatekeeping, but a ratings system is the best approach we've come up with so far for striking a balance between openness and protection. Movies are already rated in many countries and according to many different standards. Ticket takers are the cinema's gatekeepers, admitting only adults to adult shows. In the United States, where a law will soon require new televisions to be equipped with a so-called "V-chip" to allow parents to block unsuitable shows, the TV networks are formulating a ratings system of their own. The software industry is beginning to label and rate content for CD-ROM games too.

Once they mature, the ratings systems for Web sites will work most of the time, and they will help avoid what could be a real threat to the future of interactive networks—heavy-handed government interference that goes so far it jeopardizes the very openness that makes the Internet valuable.

Despite these issues and the other occasional problems posed by electronic interactions, my enthusiasm for the Internet remains boundless. Information technology is already touching lives deeply, as evidenced by a piece of electronic mail a reader of my newspaper column sent me: "Mr. Gates, I am a poet who has dyslexia, which basically means I cannot spell worth a damn, and I would never have any hope of getting my poetry or my novels published if not for this computer Spellcheck. I may fail as a writer, but thanks to you I will succeed or fail because of my talent, or a lack of talent, and not because of my disability."

We're watching something historic happen, and it will affect the world seismically, rocking us as the discovery of the scientific method, the invention of printing, and the arrival of the Industrial Age did. If the Internet could increase the understanding that citizens of one country have about the citizens of their neighboring countries and thereby reduce international tensions, that in and of itself would be enough to justify everything it cost. If the interactive network were used only by scientists and it enabled them to collaborate more effectively to find cures for the still-incurable diseases, that alone would be sufficient. If the net were only for kids, so that they could pursue their interests in and out of the classroom, that by itself would transform the human condition. The Internet won't solve every problem, but I'm convinced that it will be a positive force in many areas.

The broadband interactive network will evolve from the Internet, but not according to a preordained plan. There will be setbacks and unanticipated glitches. Some people will seize on the setbacks to proclaim that the Internet never really was more than hype. But the early failures will just be learning experiences. The Internet is terribly important.

Big changes used to take generations or centuries. This one won't take that long, but it won't happen overnight either. Within a decade we'll see widespread effects of the communications revolution, and within twenty years virtually everything I've talked about in this book will be broadly available in developed countries and in businesses and schools in developing countries. The hardware will be installed. The software will be friendly and ready to serve. Then it will be just a matter of what people decide to do with it—which is to say, what applications they end up using.

You'll know that the network has become part of your life when you begin to resent it if information isn't available to you via the network. One day you'll be hunting for the repair manual for your bicycle and you'll be annoyed that the manual is a paper document you could misplace. You'll wish it were a Web page with animated illustrations and a video tutorial, always available on the network.

The Internet will draw us together, if that's what we choose, or let us scatter ourselves into a million mediated communities. Above all, and in countless new ways, the interactive network will give us choices that can put us in touch with information, entertainment, and each other.

I think Antoine de Saint-Exupéry, who wrote so eloquently about how people came to think of railroad locomotives and other forms of technology as friendly, would applaud the coming of the Internet and dismiss as backward-looking dreamers the people who resist it. Fifty years ago Saint-Exupéry wrote: "Transport of the mails, transport of the human voice, transport of flickering pictures—in this century as in others our highest accomplishments still have the single aim of bringing men together. Do our dreamers hold that the invention of writing, of printing, of the sailing ship, degraded the human spirit?"

The information highway will lead to many destinations. I've enjoyed speculating about some of them in this book. Doubtless I've made some foolish predictions, but I hope not too many. In any case, I'm excited to be on the journey.

AFTERWORD

People often overestimate what will happen in the next two years and underestimate what will happen in ten.

I'm guilty of this myself. In 1983 I demonstrated a prototype of Windows. In 1986 Microsoft hosted our first CD-ROM conference. In each case I predicted that the technology would become important within two or three years. I was wrong, too optimistic in the short term. If anything, though, I underestimated the long-term importance of these innovations. By 1993 Windows was running on tens of millions of computers. By 1996 CD-ROM drives were standard equipment on new PCs.

In 1994 I thought that millions of Americans might be connected to broadband interactive networks as early as 1997. I wasn't alone in my optimism. Many companies expected broadband networks to deliver movies-on-demand and other entertainment into residential neighborhoods within only a few years, and telephone and cable companies promised heavy investments. Cover stories in major magazines speculated on the impending arrival of interactive television. But as we know, instead of millions of people interacting on broadband networks, what we actually got in the short term was millions of people interacting on the Internet. That doesn't repudiate the vision of a great "information highway." It's a different path leading to the same end point. By the tenth anniversary of information highway mania, the Internet will deliver the full highway we envisioned.

As I said in chapter 9, the greatest benefit of this communications revolution will be using interactive technology for learning, both inside and outside the classroom. I haven't forgotten what a difference the grant for computer time from the Mothers' Club at Lakeside School made in my life. My share of the revenues from this book goes to support the work of teachers who are incorporating computers into their classrooms. Through the National Foundation for the Improvement of Education in the United States and comparable organizations throughout the world, the funds will help teachers create opportunities for students.

I was lucky to go to a school where students had computer access, and my luck has held up. I believe I have the best job in the world. I work hard at it because I find it exciting, but my success comes from good luck too. My friend Warren Buffett, who's often called the world's greatest investor, talks about how grateful he is to live at a time when his particular talents are valuable. Warren says if he'd been born a few thousand years ago, he'd probably have been some animal's lunch. But he was born into an age that has a stock market and rewards Warren for his unique understanding of the market. Football stars should feel grateful too, Warren says. "There just happens to be a game where it turns out that a guy who can kick a ball with a funny shape through goal posts a fair percentage of the time can make millions of dollars a year," he says. I was born at the right place and time too.

But as they say in the ads for mutual funds and other financial investments, past performance is no guarantee of future results. My focus is on keeping Microsoft in the forefront through constant renewal. There are no assurances that I'll succeed.

It's a little scary to contemplate the historical truism that there's never been a leader from one computer technology era who was also a leader in the next. The fact that Microsoft has been a leader in the PC era should mean that we won't be a leader in the communications era. I'd like to defy that tradition. I still think that the tendency for successful companies to fail to innovate is just that: a tendency. If you're too focused on your current business, it's hard to look ahead and even harder to make the changes you need to.

It's easy to write the article "Microsoft Obsolete," and journalists have been doing that for more than a decade. Today Microsoft faces extraordinary opportunities as well as challenges, but it's been that way throughout the last twenty years.

The challenges have been invigorating. Between 1976 and 1979, when no one was paying much attention to tiny Microsoft, we had to prove that the personal computer was more than a toy.

With our MS-DOS partnership with IBM in 1981, Microsoft got a lot more attention. Some pundits wrote us off. They said it was just a lucky fluke that IBM had called on us to develop an operating system for the original IBM PC, and they predicted that our giant partner would drop us. During the 1980s IBM undertook a lot of projects whose goal was to eliminate any dependence on Microsoft. It was only by staying ahead of IBM technically that we maintained the relationship.

In the early 1980s many software companies targeted MS-DOS by trying to either clone it or improve on it. Digital Research, which had been the leading maker of operating systems for PCs until Microsoft entered the market, fought back with various versions of its operating system. AT&T thought its multitasking UNIX operating system was the answer, and there was rampant speculation that IBM would license it. "The previous stars—Digital Research and Microsoft—may soon find themselves playing cameo roles as AT&T and IBM take center stage," the trade journal *Computerworld* commented early in 1984. Another journal, *Datamation*, wasn't optimistic about Microsoft's future either: "What one industry analyst calls the 'irresistible tide' of AT&T's Unix now threatens to engulf the current microcomputer operating system standard, MS-DOS."

In 1983 Microsoft started work on Windows, our homegrown effort to enhance MS-DOS with a multi-windowing environment. Lots of companies had competitive strategies. Makers of the pioneering spreadsheet VisiCalc announced a graphical interface for the PC called VisiOn. The next year Digital Research announced GEM (Graphics Environment Manager), its entry in the graphical-computing sweepstakes. In 1984

Apple released the Macintosh, whose interface—like that of Windows, VisiOn, and GEM—had been inspired by pioneering work on graphical interfaces at Xerox PARC.

About this time IBM announced TopView, its own effort to add a layer on top of MS-DOS that would allow people to run different programs in different windows. TopView wasn't graphical and it had technical weaknesses, but it was a threat to Microsoft Windows because it came from IBM. *PC Magazine* asked Esther Dyson, a smart student of the industry, to predict which would be the big winner, Microsoft with Windows or IBM with TopView. "The one-line answer is IBM," Dyson said.

There were other threats in the late 1980s. The company called Next, which Steve Jobs founded after leaving Apple, developed graphical software that was a credible competitor to Windows. There were several attempts to unify UNIX into a single standard that could compete with Windows. As I mentioned in chapter 3, the most prominent of these was the Open Software Foundation, formed by several big companies including Hewlett-Packard, DEC, and even our ambivalent partner, IBM.

IBM's ambivalence lasted for years, and it unnerved me. We were always willing to do whatever IBM wanted, but eventually IBM kicked us out of the partnership. IBM decided it could write operating systems without Microsoft's help, and the separation agreement let IBM keep the name "OS/2" and the joint work the two companies had done on OS/2. It was the scariest threat to Microsoft's future ever. We had to compete against the largest company in the computer industry, which was fighting us with operating system software we had helped develop.

In response to the success of Windows, Apple and IBM temporarily set aside their rivalry in 1991 and joined forces to create Taligent and Kaleida Labs. Taligent's mission was to produce an operating system that would eclipse Windows. Kaleida's was to demonstrate market leadership in multimedia technology. Journalists took these initiatives seriously. "The joint venture between International Business Machines Co. and Apple Computer Inc., which aims to challenge Microsoft Corp. for leadership in computer operating systems, is ahead of schedule and vying aggressively for the allegiance of leading applications," the *Wall Street Journal* reported early in 1993. But Taligent and Kaleida achieved little success in the marketplace, and by 1995 Apple and IBM had shut them down.

When I became a big believer in pen-based computing in 1991, my strategy was to enhance Windows to meet the needs of people who wanted to write instead of type instructions to their computers. But not everybody thought that enhancing an existing operating system was the best approach to pen-based computing. A pioneer in the field, a company called Go, persuaded several companies to develop applications for an entirely new pen-based operating system. The difference in approach between Microsoft and Go came down to this: When you have a major shift in the way people use computers, should you adapt existing software to the new realities or start again from scratch? The question, which is relevant in the Internet era too, didn't get answered because neither pen-based effort succeeded. Handwriting-recognition software just wasn't good enough.

There were numerous other challenges to Microsoft. More than once a number of companies banded together to promote technical standards contrary to Microsoft's. At one point Sun Microsystems tried to persuade PC makers to produce computers that ran a Sun operating system. Some people expected an object-oriented scripting language from General Magic to draw users away from Windows. Wanting to bolster its software capability, IBM bought Lotus in 1995. This gave IBM Lotus Notes, one of the "middleware" software products I described in chapter 11.

Over the years some of the challenges to Microsoft have been played up in the press. Others have been all but ignored. But challenges are always there, and my job wouldn't be as much fun if they weren't. The enthusiasm we generate around Microsoft when it's time to meet a major new competitive initiative is awesome. There aren't many boring days.

Now some people wonder if Microsoft has met its match in the Internet.

The Internet makes great software more rather than less valuable. The ideal circumstance for us would be a world of almost-free PCs and almost-free infinite-bandwidth communication. Then all of the commercial value would be in the software, including content. Actually, computing and communications are moving in that direction, but slowly. There's far more opportunity than threat in the Internet.

Could we blow it? Of course. No one's future is assured. But we're moving forward, bringing the worlds of computing and communications together, innovating rapidly. Faced with the prospect of a major shift in the way people use computers, we're quickly adapting the software tens of millions of people already use, to meet new needs, while some of our competitors are building a new approach from scratch. I'm betting on Windows again.

Windows is not invincible, and neither is Microsoft. Someday Microsoft will die. Will it be a year from now or fifty years from now? Somebody at a conference asked me, "What operating system will replace Windows when it dies?" I said, "Tell me where I'm going to die, and I'll make sure not to go there."

When people predict that the Internet will do us in, they often point to the way IBM lost its leadership position. But our situation is somewhat different. IBM wasn't paying enough attention to the trends and technologies that ended up being important. Microsoft is focused intently on the Internet. If it turns out that we don't adapt well to the realities of the communications revolution, it will be because we misexecuted, not because we focused on the wrong objective.

I don't mind a few doubts about Microsoft's future. It's healthy to work in an environment where you know that commentators and competitors are watching and wondering whether you can succeed. It forces everyone at Microsoft to ask, "How well can we do?" Smart, competitive employees respond to that challenge.

For me, a big part of the fun has always been to hire and work with smart people. I enjoy learning from them. Some of the people we're hiring now are a lot younger than I am. I envy them for having grown up with better computers than I had. They're extraordinarily talented, and they'll contribute to new visions of what Microsoft can achieve. If Microsoft can combine their ideas with the contributions of our talented veterans and keep listening carefully to customers, we have a chance to continue to lead the way.

I believe more than ever that this is a great time to be alive. The dawn of the Information Age offers the best chance the world has seen to start new companies and make advances in medicine and the other sciences that improve the quality of life—or less grandly, merely to understand what is happening around us and stay in touch with families and friends, no matter where they are.

The communications revolution is empowering, but it's not a cure-all. It's important that both the good and the bad implications of technological advances be discussed broadly so that whole societies, rather than just technologists, guide our course.

Now it's back to you. I explained in the foreword that I wrote the book to help get a dialogue started and to call attention to opportunities and issues that individuals, companies, and nations will face. I hope that after reading it you share my optimism and that you'll join the discussion of how we all should help shape the future.

INDEX

and operating systems, 66, 69, 274–76
for PC architecture, 65
and PCs, 15–17, 263, 276, 279
of phone and cable companies, 264–65, 266
and scalable architecture, 42
smoke-signal, 25–26
and software, 48–49, 69
for Web browsers, 273–74, 276–77
compression, 33, 77, 304
CompuServe, 120, 143, 278
computers:
access to, 15, 291–92
applications for, 52, 55
binary expression in, 28, 31
as business tools, 154–55
capacity of, 32–33
control over, 2, 11, 93
costs of, 20, 36, 39, 213, 277
dedicated vs. general-purpose, 40–41, 77–78
early hardware, 11–16, 24, 28–30
early software, 12, 14, 15–18
growth of, 34–36
as home centers, 246
humanized, 93–95
and intelligence, 289–91
interactive, 58
kids and, 1–2, 212, 224, 294
laptop, 37, 47, 80–81
mainframe, 41–42, 56, 61, 62–64
networked, 276–77
notebook, 81
obsolescence in, 14–16
operating systems for, 41, 53–54
pen-based, 85–86
personal, see personal computers
and scalable architecture, 42
in schools, 214–15, 222–29, 292
sizes of, 12, 36–37
connectivity, 110, 261
control:
over computers, 2, 11, 93
governments and, 11, 269, 280–82, 298, 309–10, 313
in individualized learning, 217–18, 225–27
of information, 76, 269
intelligence and, 290
ratings for, 244
speed bumps for, 300
and standards, 62, 67, 281–82
copper wire, 34, 37, 104, 105, 106
copyrights, 46, 198–99, 201, 202
Corbis, 257–58
Cornell University:
CU-SEE-ME, 169
Internet use, x
critical mass, x, xi
Costa Rica, 270
cryptography, 21–22, 95–100, 301–2, 306–7; see also encryption
CU-SEE-ME, 169

data, see information
dBASE, 75
decentralization, 153, 163, 178–79, 284, 296–97
Deep Blue, 289–90
desktop publishing, 75, 139, 146–47
Diffie, Whitfield, 98

Digital Equipment Corporation (DEC):
and AltaVista, 87
and diversification, 279
in PC market, 56
PDP-1, 39–40
PDP-8, 11–12, 14, 39
PDP-11, 49
VAX, 42
digital information, 5, 23, 25, 27, 37, 129
appliances for, 73–74
architecture for exchange of, 109
on CDs, 32
modems and, 104–5
digital money, 81–82, 301–2
Digital Research, 54
directories (folders), 167, 274–75
DIRECTV, 127, 271
distribution friction, 138, 139, 141
documented life, 303–6
documents:
collaboration on, 131–32, 153, 168–70, 174
distribution of, 136–37
editable, 131
electronic, 130–37
interactive, 135–36
locating, 165
multimedia, 146, 152
paper, 129–30, 133, 136, 146
use of term, 130
Dominican Republic, 270
DOS (disk operating system), 41
MS-DOS, 54, 57, 58, 61, 62, 68, 69
PC-DOS, 54, 56
Drucker, Peter, 227
dumb terminals, see Internet terminals
DVD (digital video disk), 145

e-books, 130
e-World, 143
Eagle, 56, 263
Eckert, J. Presper, 28
EDI (Electronic Document Interchange), 156–57
Edison, Thomas, 74
education, 208–35, 317; see also learning
business investments in, 232–33
and change, 210–11, 212, 213, 231, 235, 317
at college level, 213–14, 221–22, 226
in 1890s, 211–12
homeschooling, 213
individual pace in, 218
individual responsibility for, 288–89
individual style in, 217–19
interactive, 216–17, 219–21
learning circles in, 229
mass-customized, 218
multimedia materials in, 234–35
PCs in, 209–11, 212–13, 217
policies, 231–32
problem-solving skills in, 288
and research, 222–23, 226
roles of teachers in, see teachers
self-assessment in, 226–27
simulations in, 229–30
software for, 212–13, 218–20, 221
of workforce, 213, 227
and World Wide Web, 221–22, 223
Eisenstein, Sergei, 152

- ▶

---------------------------->

- ▶